**BOCCONI
UNIVERSITY
PRESS**

Oreste Pollicino · Francesca Aurora Sacchi
Noemi Conditi

IS AI THE PERFECT DOCTOR?

Artificial Intelligence's Global Impact
on the Legal and Policy Boundaries in Healthcare

Preface by **Gianmario Verona**

Cover: Cristina Bernasconi, Milano
Typesetting: Stefania Gerosa, Milano

Copyright © 2025 Bocconi University Press
EGEA S.p.A.

EGEA S.p.A.
Via Salasco, 5 - 20136 Milano
Tel. 02/5836.5751 – Fax 02/5836.5753
egea.edizioni@unibocconi.it – www.egeaeditore.it

First edition: August 2025

ISBN Domestic Edition	979-12-80623-32-4
ISBN International Edition	979-12-81627-65-9
ISBN Digital International Edition	979-12-81627-66-6
ISBN Epub Edition	979-12-229-8048-5

Table of contents

PART III
REGULATING ARTIFICIAL INTELLIGENCE
WORLDWIDE: DIFFERENT APPROACHES FOR
ADDRESSING DIFFERENT NEEDS AND PRIORITIES

Preface

by Gianmario Verona[1]

Paraphrasing the influential 2014 book by Erik Brynjolfsson and Andrew McAfee, we are now entering the "third machine age"—an era defined by the rise of Artificial Intelligence (AI). The first age of technological transformation was marked by the advent of computers and digital devices, which, propelled by Moore's Law, became progressively smaller and more powerful. Then came the internet, connecting these devices and unlocking new dimensions of communication and information sharing. Today, we are transitioning into a new phase, in which machine learning and deep learning technologies allow us to harness both small and big data in transformative ways.

Until recently, AI was often regarded as a technology of the future— so much so that it inspired the quip: "AI is any technology that doesn't work yet." That future, however, has arrived. AI is no longer a speculative concept limited to visionaries; it is now a powerful and dynamic force reshaping every industry—and, more importantly, every aspect of our lives.

The launch of OpenAI's ChatGPT, followed by the R&D race among key global competitors, has simplified AI's interface and made advanced tools accessible to all. As Satya Nadella, CEO of Microsoft, aptly noted at a recent World Economic Forum panel, if the internet put information at our fingertips, AI puts skills at our fingertips. Nowhere are the stakes higher – or the potential greater – than in medicine and the life sciences. From research to clinical practice, digital transfromation and AI are reshaping the boundaries of what healthcare systems can deliver.

This book begins with the premise that AI's transformative impact will depend not only on technological advancement but also on the strength of the legal, ethical, and policy frameworks that support it. The authors pro-

---

[1] Gianmario Verona, Past Rector, is the Fondazione Romeo and Enrica Invernizzi Professor of Innovation Management at Bocconi University and President of Human Technopole Foundation.

vide a rigorous analysis of the key challenges raised by the integration of AI in healthcare. They start by clarifying definitions and exploring practical applications, then delve into the increasingly complex legal and policy landscape. The final section offers a global perspective, comparing how different jurisdictions—with their own regulatory cultures and strategic priorities—are responding to this rapidly evolving frontier.

What makes this volume particularly compelling and potentially interesting for readers of different fields is its interdisciplinary approach. The authors draw on jurisprudence, public policy, bioethics, and health systems research to examine how societies can govern AI in a way that fosters innovation while safeguarding equity, transparency, and accountability. This is, in my view, the only possibile venue to make a complex technology viable to a complex field like medicine.

The European perspective is presented with particular nuance, highlighting the EU's ambitious effort to build a comprehensive digital health governance framework. Grounded in initiatives like the AI Act and the European Health Data Space (EHDS), this vision aims to promote a human-centric approach that prioritizes human rights, privacy, explainability, and fairness. At the same time, the book explores the United States' approach, where emphasis is placed on post-market accountability and market-driven innovation. This comparative analysis allows readers to appreciate the structural, cultural, and philosophical differences between the two systems—and to consider what a coherent global approach to AI governance might require.

What sets this volume apart is its ability to marry doctrinal precision with practical insight. It not only maps current regulatory tools but also starts tackling deeper normative questions: How can we ensure that AI supports rather than replaces clinical judgment? How do we promote fairness and inclusivity in data-driven medicine? And how do we build resilient systems that anticipate—rather than merely react to—ethical and legal challenges?

The authors wisely recognize that these issues are not just technological or legal—they are deeply human. As AI systems grow in complexity, the patient–clinician relationship remains central to compassionate and effective care. Innovations such as predictive analytics and autonomous decision-making tools must not obscure the fact that healthcare is inherently relational, built on trust, empathy, and shared understanding. These technologies are not mere enhancements; they fundamentally reshape how we think about diagnosis, consent, safety, and accountability.

In this way, the book contributes meaningfully to a broader and urgently needed conversation. It calls on regulators, developers, clinicians, and scholars to engage with AI not merely as a tool, but as a force for system-wide transformation—one that demands new skills, new safeguards, and new ways of thinking. In the European context especially, where health is seen not only as an individual right but as a collective good, there is a unique opportunity to shape a regulatory model that is forward-looking, ethically grounded, and globally influential.

Throughout my career, I have believed in the power of research and policy to drive innovation that is inclusive, responsible, and impactful. This volume embodies that same conviction and hence it represents a resource for anyone seeking to navigate the evolving landscape of AI in medicine—not with fear or uncritical enthusiasm, but with the intellectual seriousness and vision that this moment demands.

Introduction

The promises of Artificial Intelligence to revolutionize the way in which diagnoses, treatments, and patient care are provided have now reached almost every conference, roundtable, and paper on the future of healthcare, as well as everyone's lips in discourses on the matter. Indeed, both private and public investments in the development of innovative technological solutions to be deployed for the provision of healthcare have drastically increased in the past few years. To name but a few examples, the United States, through agencies such as the National Institutes of Health (NIH) and the Department of Health and Human Services (HHS), has funded AI-driven projects aimed at improving diagnostics, drug discovery, and personalized medicine. Similarly, the European Union has launched large-scale initiatives, such as the Horizon Europe program, which includes significant funding for AI applications in healthcare, with a total amount of 297 projects funded in 2024 since 2021.[1] In China, AI development is a strategic priority, with the government investing heavily in AI-powered medical technologies as part of its broader push for leadership in AI innovation. Private investments have also surged, with major technology companies dedicating resources to research projects on the matter.

As a consequence, new AI systems to be implemented in healthcare include, to name just a few examples, AI-powered imaging systems that detect cancerous tumors with remarkable accuracy, predictive analytics that help anticipate disease outbreaks, tools, and software to help better administer hospital and healthcare facilities, and the list seems endless and meant to include unpredictable results. Therefore, it goes without saying that their integration into healthcare continues to generate significant excitement among stakeholders, healthcare professionals, and patients.

---

[1] Raffaele Guerini, «And the winners are: Horizon Europe funding for artificial intelligence is surging, a Scienc|Business analysis finds», *Science Business*, March 21, 2024.

At the same time, however, concerns about the need to carefully consider the complex set of legal and ethical challenges posed by the design, development, and deployment of such technology have long gained momentum, especially in the field of healthcare, where core fundamental rights and values are involved and potentially endangered. Indeed, the introduction of such technology without carefully considering its impact on patients and professionals can have adverse and unprecedented consequences that only recent studies have begun to evaluate. There is vast evidence for the need for a careful approach, which includes for instance that it has been proven ChatGPT performs incorrect diagnoses in more than 8 in 10 pediatric case studies,[2] or that AI systems can give biased results, thus "worsen[ing] healthcare disparities."[3]

However, technological advancements in the field of AI, particularly in healthcare, as well as the implementation of AI systems by healthcare professionals, have so far occurred almost in a legal vacuum, given the complete absence, until very recently, of legal norms specifically aimed at regulating this area. Indeed, traditional legal principles that govern product design and manufacturing, medical malpractice, informed consent, and many more are often not suitable for the purpose, considering the unique characteristics of AI systems, particularly their ability to act as both autonomous decision-makers and decision-support tools, as well as to introduce new risks concerning data privacy and cybersecurity. Therefore, around the globe, the need for both interpreting and adapting existing norms and enacting new ad hoc regulations has emerged to guide the development of such technology. At the same time, however, providing rules and guidance while fostering innovation necessitates a careful balancing act between competing interests and shall be deeply rooted in shared ethical principles. These rules should ideally be as shared as possible among different jurisdictions precisely because of the fact that fundamental rights of people and values of mankind are potentially triggered by considering that AI applications in healthcare are not confined to national borders, data is often shared across jurisdictions, and AI models may be trained on global datasets. Consequently, harmonizing legal frameworks across countries is essential to ensure an ethical deployment of AI while facilitating innovation. The role of global regulatory bodies in establishing international standards for AI in healthcare will be crucial in shaping the future of AI governance.

---

[2] Joseph Choi, «ChatGPT incorrectly diagnosed more than 8 in 10 pediatric case studies, research finds», *The Hill,* March 1, 2024.
[3] Mike Miliard, «Yale study shows how AI bias worsens healthcare disparities», *healthcare IT News,* November 25, 2024.

This volume precisely seeks to provide a comprehensive analysis and overview of AI in the healthcare sector, as well as an overview of the current ethical, policy, and legal debate on the topic and insights on future developments for policymakers, legal practitioners, healthcare professionals, and scholars. Moreover, it will serve as a roadmap for navigating the complex landscape of AI in healthcare. By fostering an informed and nuanced discussion, this book aspires to contribute to the broader discourse on AI governance and ensure that technological advancements align with legal and ethical principles.

To this end, Part I aims to define the perimeter of the subsequent discussion in terms of definitions and generally describe the vast range of possible applications of such technology in healthcare as the basis for the discussion in the pages that follow and to provide an overview of the state of the art.

In Chapter 2, we will delve specifically into the most pressing ethical and legal concerns on the matter in order to provide a general overview of the theoretical issues that should be addressed for ethically sound development and deployment of AI systems and the enactment of appropriate norms. By engaging with these issues, stakeholders can work toward a future in which AI-driven healthcare not only improves patient outcomes but does so in a manner that is legally sound, ethically responsible, and socially just.

Chapter 3 navigates the complex policy landscape on AI in healthcare, highlighting the difficulties policymakers face in shaping national and international regulations. The unique needs and characteristics of AI in this sector are taken into consideration and the importance of ensuring a delicate balance between research freedom and innovation boosting while protecting fundamental rights and patient safety is highlighted.

Part III is entirely devoted to providing a footprint of the current legal framework applicable to AI worldwide, with a focus on the norms and regulations of the highest interest in the healthcare sector. To this end, in Chapter 4, we analyze thoroughly the European framework, and therefore the newly enacted Regulation 2024/1689 (AI Act) and its interplay with the medical device regulation and the norms applicable to the processing of personal data. Chapter 5, on the other hand, is devoted to analyze the regulatory approach adopted in the US. It reviews the key acts and initiatives on AI regulation undertaken from the Obama administration onward, up to the second Trump administration, taking into account the different aims pursued by each President. The chapter covers also initiatives by the FDA and other federal agencies, devoting then some space to state-level efforts. Finally, in Chapter 6, we conduct the same analysis in other relevant legislations around the globe, starting from the G7 countries to then move to other pioneer countries leading in AI regulation and international policy shaping.

We are deeply grateful and honored to have Prof. Gianmario Verona contribute to this book and sincerely thank him for writing the preface. His words enrich this volume with an insightful and valuable perspective.

Oreste Pollicino is the author of Part 1 and Chapter 6. Francesca Aurora Sacchi is the author of Chapters 3 and 5. Noemi Conditi is the author of Chapters 2 and 4.

PART I
DEFINITIONS AND APPLICATIONS OF ARTIFICIAL INTELLIGENCE IN HEALTHCARE

1 Artificial Intelligence in Healthcare

1.1 Introduction

The impact of AI in the healthcare sector has been and will continue to be tremendous, with the potential to revolutionize the way we have always conceptualized the provision of healthcare. However, at the same time, the implementation of AI systems in healthcare is a highly complex field from the technical, ethical, political, and regulatory points of view. The absence of common standards, definitions of the major concepts to be addressed, and (up to very recently) comprehensive regulations on its main issues further exacerbate the already existing controversies on the topic. The fact that the implementation of AI in this sector may possibly have consequences that exceed the boundaries of States and would theoretically require an international regulation as uniform as possible makes the discussion and the search for a solution all the more complicated.

As a consequence, and with the ambitious goal of providing clarity for a discussion on AI in healthcare, in this first chapter, we intend to present a general overview of the topic by first of all defining the main technical concepts to be included in the following analysis and which are also commonly referred to in the field. Indeed, considering the inherent multidisciplinary nature of the two main intertwined topics, AI on the one side and healthcare on the other, any analysis of the current regulatory and policy landscape cannot be conducted in the absence of consensus on the meaning of its main elements. Subsequently, we will discuss some of the primary concrete areas where AI can be effectively implemented, aiming to portray a comprehensive yet concise picture of the current situation and hopefully lay the foundations for the discussion that follows.

1.2 Defining the perimeter of the analysis: artificial intelligence

The discussion should commence with an effort to define what AI is, which however remains perhaps (one of) the most complex legal task(s). Indeed, notwithstanding the strong and revolutionizing impact of AI technologies in everyday life, and especially healthcare, since the birth of the term at the Dartmouth Conference in 1956, numerous attempts have been made at defining «artificial intelligence» and at reaching consensus on its conceptual and normative characteristics. Indeed, as notorious, there is no agreed-upon definition of artificial intelligence (AI) in general, nor is there a specific one for AI in healthcare. The many suggestions that have been made have alternatively adopted a technical, ethical, or legal perspective, but in none of the mentioned fields, is there agreement or consensus on a specific definition of the term «AI» or a set of identifying elements.

The challenges that arise from trying to identify a legal definition of AI are substantially analogous to those more generally of regulating technology and science,[1] mostly related to the difficulty of fixing in time the identifying elements of a continuing evolving concept from a technical point of view, and of defining in legal terms what pertains to a different field (being that computer science, science, medicine etc.).

Up to recent years, most of the proposed definitions or tools to help qualify a technology or a system as AI have shared the common trait of referring to human intelligence as what the system should aim at replicating or emulating.[2] Already the first attempt at grasping the concept of AI made by Turing in 1950 focused on assessing whether a computer was able to imitate human behaviors. Indeed, according to this method, the so-called Imitation Game, a computer should qualify as AI if it is able to exhibit human-like intelligence and behavior. This general tendency is complemented by the philosophical debate around the definition of *intelligence* and *human intelligence*, which could serve as the preconceptual conditions for reaching the mentioned goal.

Indeed, in the late 1990s, AI was defined as «the science and engineering of making *intelligent machines*, through algorithms or a set of rules, which the machine follows *to mimic human cognitive functions*, such as learning and

---

[1] Hannah Ruschemeier, «AI as a challenge for legal regulation—the scope of application of the Artificial Intelligence Act proposal», *ERA Forum*, 23, 2023, pp. 361–376.

[2] Tijs Vandemeulebroucke, «The ethics of artificial intelligence systems in healthcare and medicine: from a local to a global perspective, and back», *Pflugers Archiv: European Journal of Physiology*, 2024, pp. 1–11. The authors make here reference, for instance, to the definitions provided by Secinaro and colleagues («artificial intelligence [...] generally applies to computational technologies that emulate mechanisms assisted by human intelligence, such as thought, deep learning, adaptation, engagement, and sensory understanding») or by Morley and colleagues («AI [is] an umbrella term for a range of techniques that can be used to make machines complete tasks in a way that would be considered intelligence were they to be completed by a human»). Silvana Secinaro *et al.*, «The role of artificial intelligence in healthcare: a structured literature review», *BMC Medical Informatics and Decision Making*, 21(125), 2021, pp. 1–23; Núria Vallès-Peris, Miquel Domènech, «The ethics of AI in healthcare: a mapping review», *AI & Society: Knowledge, Culture and Communication*, 38, 2023, pp. 1685–1695.

problem solving»[3] (emphasis added). Along this line moved also the first institutional definitions of the term. In this regard, reference can be made for instance to the one proposed by the European Commission in its 2018 Communication on AI[4] in which AI is explicitly qualified as «systems that display *intelligent behaviour* by analysing their environment and taking actions—*with some degree of autonomy*» (emphasis added).[5] Similarly, the Organisation for Economic Co-operation and Development (OECD) in the first definition provided in its Recommendation on Artificial Intelligence of 2019, subsequently amended with substantial changes in 2024, stated that AI systems are «machine-based» systems, designed to operate «with varying levels of autonomy» and that can make predictions, recommendations, or decisions, «for a given set of human defined objectives».[6]

However, issues with the mentioned approach gradually started to arise and relate in particular to the risks of «anthropomorphising artificial intelligence»,[7] because of the frequent reference to *intelligence,* a natural and ontological *human* characteristic,[8] and to *autonomy* in the decision-making process, usually referred to as the human ability to somehow adapt to a given environment. While conceptual borrowing is a fairly frequent practice and most of the time a necessity for the purpose of regulating matters that pertain to different fields, it is a practice that runs the risk of borrowing concepts that carry a baggage of meanings and implications that may «confuse

---

[3] John McCarthy, *What is artificial intelligence?*, 1998.

[4] Communication from the Commission to the European Parliament, the European Council, the Council, the European Economic and Social Committee and the Committee of the Regions on Artificial Intelligence for Europe, Brussels, 25.4.2018 COM(2018) 237 final.

[5] The full definition reads as follows: «Artificial intelligence (AI) refers to systems that display intelligent behaviour by analysing their environment and taking actions—with some degree of autonomy—to achieve specific goals. AI-based systems can be purely software-based, acting in the virtual world (e.g. voice assistants, image analysis software, search engines, speech and face recognition systems) or AI can be embedded in hardware devices (e.g. advanced robots, autonomous cars, drones or Internet of Things applications)», The European Commission's high-level expert group on AI, *A definition of AI: Main capabilities and scientific disciplines*, 2018.

[6] «An AI system is a machine-based system that can, for a given set of human defined objectives, make predictions, recommendations, or decisions influencing real or virtual environments. AI systems are designed to operate with varying levels of autonomy».

[7] Luciano Floridi, and Anna C. Nobre, «Anthropomorphising machines and computerising minds: the crosswiring of languages between artificial intelligence and brain & cognitive sciences», *Minds and Machines*, 34(5), 2024, pp. 1–9; Giusella Finocchiaro, «The regulation of artificial intelligence», *AI & Society*, 39, 2024, pp. 1961–1968.

[8] Pei Wang, «On defining artificial intelligence», *Journal of Artificial General Intelligence*, 10, 2019, pp. 1–37.

or misguide.»[9] In particular, in this specific context, linking by definition AI to human capabilities may grant AI systems with «unwarranted biological and cognitive properties that taint its understanding in society» and ultimately have «negative consequences»,[10] such as for instance leading to sci-fi fears of AI supremacy over humans and the like.

For this reason, it has been at times suggested to eliminate any reference to «intelligence»,[11] or to any human capabilities altogether,[12] or at least to make clear that any such reference is to be used only as a metaphor[13] to stress the importance of assessing whether a system can behave «as *if it* were intelligent».[14] This would nonetheless imply the necessity to constantly evaluate both benefits and limits of this legal artifice when choosing what to qualify as AI and, consequently, to be regulated in theory and implemented in practice.

It is worth noticing that an updated version of the mentioned European Commission's definition makes no reference to any anthropomorphic elements. Indeed, on the one hand, after having addressed the concept of «intelligence» as «vague», the HLEG provided a new definition in 2018, that does not refer to «intelligent behavior» or «degree of autonomy», but merely qualifies as AI those systems that are «designed by humans» to achieve a given goal.[15] At the same time, while the OECD in its recently updated version kept the reference to autonomy, the explanatory memorandum explicitly addresses the issue and defines the scope of the concept, by confining it to «the degree to which a system can learn or act without human involvement *following the delegation of autonomy and process automation by humans*».[16]

---

[9] Luciano Floridi, Anna C. Nobre, *op. cit.* p.3.

[10] Luciano Floridi, Anna C. Nobre, *op. cit.*

[11] See for instance Agrawal, Gans and Goldfarb, who suggested to substitute «artificial intelligence» with «prediction machine». Ajay Agrawal, Joshua Gans, and Avi Goldfarb, *Prediction machines: The simple economics of Artificial Intelligence*, Boston, Harvard Business Press, 2018.

[12] Daniel C. Dennett, «What can we do?», in John Brockman (ed. by), *Possible minds: Twenty-five ways of looking at AI*, New York, Penguin Press, 2019, pp. 41–53; Virgina Dignum, *Responsible artificial intelligence: How to develop and use AI in a responsible way*, Springer, 2019.

[13] On the role of metaphors and frames within legal and judicial language, especially in the digital era, see Alessandro Morelli, and Oreste Pollicino, «Metaphors, Judicial Frames, and Fundamental Rights in Cyberspace», *The American Journal of Comparative Law*, 68(3), 2020, pp. 616–646.

[14] Giusella Finocchiaro, *op. cit.*, p.1962.

[15] The European Commission's High-Level Expert Group on AI, *A definition of AI: Main capabilities and scientific disciplines*, 2018.

[16] «AI system autonomy (contained in both the original and the revised definition of an AI system) means the degree to which a system can learn or act without human involvement following the delegation of autonomy and process automation by humans. Human supervision can occur at any stage of the AI system lifecycle, such as

Albeit challenging, defining the perimeter of the concept of AI is fundamental from both a practical and a theoretical point of view. As for the first,
without a definition of AI there could not be agreement on how to concep-
tualize its design and research projects related to it and on the foundations
for the growth of the research community in the field.[17] Furthermore, defining AI is also paramount from a theoretical perspective, because of the
need to identify the perimeter of the discussion on the matter and any regulations to be enacted in this regard. From a strictly legal point of view, legal
working definitions are prerequisites for ensuring legal certainty,[18] because
they delimit the scope of applicability of the norms specifically enacted. In
this sense, legal definitions are context-dependent, because they should be
fit for the specific context they will be used in. In healthcare and for the
purposes of regulating AI and its use, the legal definition of AI is to be
teleologically interpreted[19] and should serve the objective of identifying the
scope of application of the norms (to be) enacted. Such a definition would
thus help differentiate between systems that would fall within the scope of
application of any regulation on the matter and should therefore comply
with the legal requirements enshrined therein and which are otherwise free
from such obligation. Considering that legal requirements on the matter
should theoretically be provided to protect the fundamental rights and interests of individuals possibly affected by AI while at the same time adequately
fostering its development and use for the benefit of society,[20] the goal of any
regulation of this activity should thus be to strike a fair balance between all
the competing interests at stake, and such a complex exercise begins with
properly defining its scope of application and thus, for our purposes, AI.

As a consequence, from an ethical perspective, it has been suggested that
when it comes to finding a definition of AI, «AI paradigms still satisfy the
classic definition provided by John McCarthy»[21] *et al.*, where intelligence

---

during AI system design, data collection and processing, development, verification,
validation, deployment, or operation and monitoring. Some AI systems can generate
outputs without these outputs being explicitly described in the AI system's objective
and without specific instructions from a human», Explanatory memorandum on the
updated OECD definition of an AI system, OECD, 2024.

[17] Pei Wang, *op. cit.*

[18] Hannah Ruschemeier, *op. cit.*

[19] Giovanni Sartor, *L'intelligenza artificiale e il diritto*, Torino, Giappichelli, 2022.

[20] In general, on the possible challenges to fundamental rights posed by AI technologies and systems, see Stéphanie Laulhé Shaelou, and Yulia Razmetaeva, «Challenges to fundamental human rights in the age of artificial intelligence systems: shaping the digital legal order while upholding Rule of Law principles and European
values», *ERA Forum 24*, 2023, pp. 567–587.

[21] Luciano Floridi, «What the near future of artificial intelligence could be», *Philosophy & Technology*, 32, 2019, pp.1–15.

remains a human feature and the focus is on the action performed by the AI system, and not on the possibility of qualifying it as *intelligent* in itself. Therefore, «reservoir of smart agency on tap» is the working definition of AI that has been suggested.[22] Differently, strictly adopting the point of view of legal theory, among other requirements a legal definition of AI in particular should be:

- *inclusive* in relation to the identified regulatory goal.[23] This require-ment entails simultaneously avoiding both an *under-inclusive* defini-tion, i.e., excluding from the scope of application of the norms aspects or uses of AI that should be included according to the regulatory goal, and an *over-inclusive* one, i.e., too broad, so that any possible software or software-based technology would fall within. Both under- and over-inclusiveness prevent the regulatory purpose from being reached, by not sufficiently protecting fundamental rights potentially affected by AI on the one hand or by imposing strict requirements for AI sys-tems that do not pose such risks (or that do not pass the threshold of seriousness) on the other. This principle, especially when applied to the supranational level of normative interventions, arises from the princi-ple of proportionality;
- *precise and comprehensive*, and therefore a definition that the interpreter is able to correctly understand, particularly one that allows for deter-mining whether a given situation falls thereunder. The aforementioned characteristics are recommended to ensure an adequate level of legal certainty;[24]
- *flexible* enough to adapt to future developments of the technology. In-deed, it would not be feasible (nor practical) to amend or modify the provided legal definition any time that a new AI would fall outside its scope of application because of a technical development;
- including, if needed, anthropomorphic references, as long as any con-fusion between human and machine characteristics is avoided, in the sense highlighted above.

Moreover, on the identified legal definition, there should ideally be general agreement, possibly at the international level, to ensure both legal certainty

---

[22] Luciano Floridi, and Josh Cowls, «A united framework of five principles for AI in society», *Harvard Data Science Review*, 1(1), 2019, pp. 1–14.

[23] Jonas Schuett, «A legal definition of AI», *SSNR Electronic Journal*, 2019, pp. 1–15.

[24] Giovanni Sartor, *op. cit.*

and legal and technical interoperability among different jurisdictions. While this goal remains ambitious, it should nonetheless be pursued to ensure an adequate level of technical interoperability among systems in different jurisdictions and a uniform level of protection of the rights and interests possibly affected by the use of these technologies. After all, the consequences of the use of AI systems, especially for the provision of healthcare, may have transnational consequences for both patients and healthcare professionals that legal norms should aim at safeguarding.

1.2.1 The European approach—Regulation 2024/1689 (AI Act)

On July 12, 2024, the Regulation 2024/1689 (so-called AI Act or AIA) was officially published in the Official Journal of the European Union what has been defined as the first set of rules on AI in the world.[25] While this might not necessarily be true, considering that for instance the US National AI Initiative Act became law three years earlier on January 1, 2021,[26] the European Union took the substantial lead in regulating AI[27] and the AI Act has nonetheless the merit of being «one of the most influential regulatory steps taken internationally».[28]

The purpose of the AI Act, as already outlined in the Proposal published on April 21' 2021, is well described in its first recital:

> to *improve the functioning of the internal market* by laying down a *uniform legal framework* in particular for the development, the placing on the market, the putting into service and the use of *artificial intelligence systems (AI systems)* in the Union, in accordance with Union values, to *promote the uptake of human centric and trustworthy artificial intelligence (AI)* while ensuring a *high level of protection of health, safety, fundamental rights* [...] and *environmental protection,* to protect against the harmful effects of AI systems in the Union, *and to support innovation.* (Emphasis added).

The explicit regulatory purpose of the AI Act is therefore (at least) twofold: It aims at providing norms to adequately protect fundamental rights possibly affected by AI, as well as the environment, while at the same time foster-

---

[25] As announced on a press release by the Council of the European Union. The same statement had already been made by the European Data Protection Supervisor about the Proposal in 2021: https://edps.europa.eu/press-publications/press-news/press-releases/2021/artificial-intelligence-act-welcomed-initiative_en

[26] Luciano Floridi, «The European legislation on AI: a brief analysis of its philosophical approach», *Philosophy & Technology,* 34, 2021, pp. 215–222.

[27] Jonas Schuett, *op. cit.*

[28] Luciano Floridi, «What the near future of artificial intelligence could be», *op. cit.*

ing the development of trustworthy, human-centric, and safe AI systems and thus improving the functioning of the internal market.[29] This complex regulatory goal and balancing exercise are achieved through the definition of product technical requirements, obligations on the economic operators or other persons somehow involved in the supply chain and life-cycle of the AI system, and powers of public supervisory authorities accordingly. As it will be further discussed in the following chapters, the European approach adopted in this regard is *risk-based*, i.e., it establishes stringent provisions for AI systems that pose greater risks to the underlying protected values, and *horizontal*, namely that provisions are enacted for the design, development and use of AI systems in general, whatever the concrete field of its application.[30] However, at times the regulation differentiates between various fields to adjust some of its provisions to the specific risks.

For the purpose of developing the Proposal, the European Commission had to define its scope of application and therefore to identify clearly what AI is.[31] Given the importance and innovative impact of this regulation on the development of AI, also worldwide, it is interesting to understand how and if the mentioned regulatory goals have been accomplished.

The initial proposal for a definition underwent substantial changes to reach its final version, which substantially replicates the one provided by the OECD.

Original version of the definition art. 3 Proposal AI Act	Final version of the definition art. 3 AI Act
«artificial intelligence system» (AI system) means *software* that is *developed with one or more of the techniques and approaches listed in Annex I* and can, for a *given set of human-defined objectives*, generate outputs such as content, predictions, recommendations, or decisions influencing *the environments they interact with*.	«AI system» means a *machine-based system* that is *designed to operate with varying levels of autonomy* and that may exhibit *adaptiveness after deployment*, and that, for *explicit or implicit objectives*, infers, from the input it receives, how to generate outputs such as predictions, content, recommendations, or decisions that can influence *physical or virtual environments*.

---

[29] The same aim is also clearly stated and further explained in the Explanatory Memorandum of the Proposal. On this, see also Daniela Messina, «La proposta di regolamento europeo in materia di Intelligenza Artificiale: verso una «discutibile» tutela individuale di tipo consumercentric nella società dominata dal «pensiero artificiale», *Rivista di diritto dei media*, 2/2022, 2022, pp. 196–231.

[30] Pietro Falletta, and Annalisa Marsano, «Intelligenza artificiale e protezione dei dati personali: il rapporto tra Regolamento europeo sull'intelligenza artificiale e GDPR», *Rivista italiana di informatica e diritto*, 1/2024, 2024, pp. 119–137.

[31] A detailed and more in-depth analysis of the structure and the provisions of the Regulation is provided in Chapter 4.

Some aspects were not subject to revisions throughout the legislative process and therefore may represent the underlying approach of the European Union to the matter. First of all, in both versions of the definition, it was decided to define «AI *systems*» instead of the more general term «AI». This choice probably stems from considering that AI is sometimes also used to refer to the field more generally,[32] rather than just to the specific technological system developed and used.

Moreover, there are no «sci-fi speculations about AI» that had appeared in previous works of the European Parliament, such as in the European Parliament resolution of 16 February 2017 with recommendations to the Commission on Civil Law Rules on Robotics, or the HLEG, in particular in the Draft Ethics Guidelines for Trustworthy AI.[33] On the contrary, the AI Act avoids any implicit or explicit reference to human-like capabilities of the system to be regulated, except for the sole reference to «varying levels of autonomy», taken here in the sense of the Explanatory Memorandum of the OECD.

Moreover, even though the AI Act adopts a risk-based approach for assessing the level of concrete regulation applicable to a given AI system, it does not define the concept of *AI systems* by referring to the risks that the latter is posing to fundamental rights. Instead, AI systems are defined in general. While the in-depth analysis of the content of the AI Act and its risk-based approach is provided in Chapter 4, it is nonetheless useful here to underline that there is in the Regulation a dichotomy between the *scope* of the AI Act, i.e., to regulate every AI system, and its *content*, which basically refers to high-risk systems exclusively. In particular, the majority of the norms of the AI Act are applicable only to high-risk AI systems, with very few of them being referred to AI systems of whatever risk classification. This led some commentators to define the definition of AI system provided for in the AI Act as an «empty shell» used for «communications purposes», because it has «no filter-function for the application of the regulation».[34]

In the intentions of the European Commission, the proposed (final) definition ought to be «future-proof»[35] and therefore flexible enough to adapt

---

[32] For instance, among many others, reference is made to McCarthy's definition «Artificial intelligence means the science and engineering of making *intelligent machines*» (emphasis added): John McCarthy, *op. cit.*

[33] Luciano Floridi, «What the near future of artificial intelligence could be», *op. cit.*

[34] Hannah Ruschemeier, *op. cit.*

[35] Explanatory memorandum—Proposal for a Regulation Of The European Parliament And Of The Council Laying Down Harmonised Rules On Artificial Intelligence (Artificial Intelligence Act) And Amending Certain Union Legislative Acts, European Commission, 21st April 2021, https://eur-lex.europa.eu/legal-content/EN/TXT/HTML/?uri=CELEX:52021PC0206.

to *machine-based systems* of the type identified as objects of the regulation, independently of future technical developments of the machines themselves. The goal of the Commission here is to contrast, as far as possible, regulatory obsolescence, which frequently affects norms intended to regulate technology (as well as science), rendered obsolete because of the continuously new instances of protections and risks that may arise from the fast technical development, and the vast and widespread use of the regulated technologies.[36]

Few changes have been made to reach the final version of the definition of AI systems. First, unlike the Proposal, the AI Act now regulates «machine-based» systems instead of the previous use of «software». Apart from the willingness to harmonize definitions across different legal documents on the matter, such as the one provided by the OECD, this choice could also derive from the need to adopt a more inclusive wording (machine-based system) that encompasses both the hardware and software parts of AI.

Moreover, reference to the specific «techniques and approaches» listed in Annex I as one of the core criteria for defining the scope of application of the norms has been removed and substituted with a more abstract description of the modes for designing a machine-based system. This way, the legislator does not intervene a priori in identifying and listing the specific techniques that, if used for a given system, make it fall under the scope of the Regulation, because potentially endangering fundamental rights, but leaves such evaluation open. This approach contributes to the future-proofing essence of the final definition. Indeed, will new techniques be developed in the future, the AI system designed and produced using them will be assessed against an abstract definition to determine whether the AI Act is applicable, without needing to amend any fixed list provided for in the Regulation itself.

Finally, the choice to substitute the characterization of the AI objectives as «human-defined», whose aim could have been to ensure the «human-in-the-loop» factor, with «explicit or implicit» seems to have added vagueness to the concept.

Overall, the definition provided for in the AI Act is not perfect, but can be considered «a good *starting* point» (emphasis added) for the design, marketing, and development of AI systems in Europe that is «ethically sound, legally acceptable, socially equitable, and environmentally sustainable».[37] At the same time, such a definition may be seen as a good *ending* point for the previous strong debate on the elements that should be included in a defi-

---

[36] Daniela Messina, *op. cit.*
[37] Luciano Floridi, «What the near future of artificial intelligence could be», *op. cit.*

nition of AI, because necessary, and those that should be excluded from it, because vague, dangerous, and confusing.

In particular, considering the above-identified necessary (but not necessarily sufficient) requirements of a legal definition of AI, art. 3 of the AI Act complies with the requirements of precision and comprehensiveness, as well as flexibility (especially considering the suppression of the strict list of techniques possibly to be used in Annex I of the AI Act). Moreover, the only human-like reference is the «level of autonomy», probably included in the same sense as the OECD's definition. However, it is sometimes questioned whether this definition of AI is adequately inclusive, and not over-inclusive, considering its aim of establishing strict rules for the design, marketing, and implementation of AI in various fields. An over-inclusive definition, as mentioned, would possibly result in a disproportionate restriction of technological developments. In this regard, we believe that an appropriate balance may have been reached precisely by the adoption of a risk-based approach that graduates normative limits and requirements according to the concrete risks posed by a given technology considered.

The definition provided for in art. 3 AI Act will serve the goal of harmonizing such a concept within the European Union, especially thanks to the choice of the regulation as a legal instrument. However, as mentioned, for an ideal level of interoperability of AI systems from both a technical and a legal point of view, ideally such a definition should be agreed upon internationally. While no current uniform definition of this kind has been approved, there is evidence of an emerging «Brussels-Washington consensus» when the European AI Act and the American Executive Order on the Safe, Secure, and Trustworthy Development and Use of Artificial Intelligence are compared.[38]

1.3 Defining the perimeter of the analysis: other relevant terminology

AI does not refer ubiquitously and universally to a single technology,[39] but it is an «umbrella term»[40] used to identify several subfields (such as, for instance, machine learning, deep learning, and natural language processing).[41]

---

[38] *Ibid.*

[39] Junaid Bajwa et al., «Artificial intelligence in healthcare: transforming the practice of medicine», *Future Healthcare Journal*, 8(2), 2021, pp. 188–194.

[40] Giuseppe Mobilio, «L'intelligenza artificiale e le regole giuridiche alla prova: il caso paradigmatico del GDPR», *Federalismi.it*, 16/2020, 2020, pp. 266–298.

[41] Peter Stone *et al.*, «Artificial intelligence and life in 2030», *One Hundred Year Study on Artificial Intelligence: Report of the 2015–2016*, Study Panel, Stanford University, 2016.

At the core of any of these fields and applications of AI (and AI itself) rests the *algorithm,* which is the fundamental unit of any technology of this kind. In itself, an algorithm is a set of precise instructions on the steps to be taken to reach a specific goal,[42] sometimes also applying learning mechanisms[43] (self-learning or adaptive algorithms, as opposed to locked algorithms[44]). Any AI system is always made (at least) of one or more algorithms, while the opposite is not true, and therefore not any system that includes an algorithm shall be qualified as AI.

At the beginning of the development of AI technologies, these algorithms were a series of instructions generated based on the knowledge of the «expert in the field»[45] on how to achieve a specific goal. With the increasing availability of data, data-driven methods were developed. This way, systems were not «externally» instructed on how to reach a certain goal, but had to deduce this process and the related set of instructions by analyzing large amounts of inputs and correlated outputs, made available by the provider.[46]

The process so described, referred to as *machine learning* (ML), is a sub-concept or branch of AI[47] that has therefore been defined as «the study of algorithms that allow computer programs to automatically improve through experience».[48] Thanks to ML, machines detect patterns, learn from mistakes, and adapt to a given environment and improve over time.[49] The input data analyzed in order to construct the algorithm are usually structured data (imaging, genetic and EP data etc.), from which the goal is to extract certain features in order to link it to a specific (set of) output.[50] ML is thus the processing of learning (*training*) models through data (*training dataset*).[51] The accuracy of the resulting system is therefore highly dependent on the

---

[42] Giuseppe Mobilio, *op. cit.*

[43] Giovanni Sartor, *op. cit.*

[44] Barry Solaiman, and I. Glenn Cohen, «A framework for health, AI and the law», in Barry Solaiman, and I. Glenn Cohen (ed. by) *The Research Handbook of AI and Healthcare*, Elgaronline, 2024, pp. 1–19.

[45] Devillé Remembrandt, Nico Sergeyssels, and Catherine Middag, «Basic concepts of AI for legal scholars», in Jan De Bruyne, and Cedric Vanleenhove (ed. by), *Artificial Intelligence and the Law*, 2021, pp. 1–22.

[46] *Ibid.*

[47] Rowland W. Pettit *et al.*, «Artificial intelligence, machine learning, and deep learning for clinical outcome prediction», *Emerging Topics in Life Sciences*, 5(6), 2021, pp. 729–745.

[48] Junaid Bajwa *et al.*, *op. cit.*

[49] Rowland W. Pettit *et al.*, *op. cit.*

[50] Fei Jiang *et al.*, «Artificial Intelligence in healthcare: past, present and future», *Stroke and Vascular Neurology*, 2, 2017, pp. 230–243.

[51] Devillé Remembrandt, *op. cit.*

quantity and quality of the input data.[52] Various categories of ML can be identified, based on the type of learning algorithm: supervised, unsupervised, and reinforced learning,[53] with ongoing research for the categories of semi-supervised, self-supervised, and multi-instance.[54]

Supervised learning occurs when the inputs are pre-catalogued (*labeled*) data, and the algorithm learns which output should be associated with it. On the other hand, in unsupervised learning, the algorithm has to autonomously infer and correlate the output to the input data[55] that are not labeled and therefore no given set of output is provided to the machine to learn from.[56] Among the other subcategories, semi-supervised is conceptualized as a combination of both supervised and unsupervised learning (and therefore some data are labeled and some others are not).[57]

Differently, in reinforced learning, the machine is not instructed on how to perform a certain task, and thus reach a given goal, but it learns either by expert demonstration or by attempting to autonomously find the optimal way to obtain the desired goal, on which positive or negative feedbacks are provided according to the correctness of the solution proposed by the machine.[58] This way, the machine can learn from its successes and mistakes.[59]

However, precisely because of the use of ML techniques, sometimes understanding why and how an AI system has reached a specific conclusion is complex, if not impossible. This phenomenon is addressed as *black-box*, i.e., when the result or outcome of a machine is not explained in a way that is understandable for humans.[60] When applied to healthcare, it can lead to *black-box medicine*,[61] when either the conclusion of the AI system cannot be explained at all (as generally for black-box AI) or in medical terms.[62]

---

[52] Shuroug A. Alowais *et al*, «Revolutionizing healthcare: the role of artificial intelligence in clinical practice», *BMC Medical Education*, 23(689), 2023, pp. 1–15.

[53] For a more in-depth explanation of the concepts, see Bertalan Meskó, and Marton Görög, «A short guide for medical professionals in the era of artificial intelligence», *NPJ Digital Medicine*, 3(126), 2020, pp. 1–8.

[54] Junaid Bajwa *et al.*, *op. cit.*

[55] Andreas Kaplan, and Michael Haenle, «Siri, Siri, in my hand: Who's the fairest in the land? On the interpretations, illustrations, and implications of artificial intelligence», *Business Horizons*, 62(1), 2019, pp. 15–25.

[56] Junaid Bajwa *et al.*, *op. cit.*

[57] Valentina Bellini *et al.*, «Understanding basic principles of Artificial Intelligence: a practical guide for intensivists», *Acta bio-medica*, Atenei Parmensis, 93(5), 2022, pp. 1–15.

[58] Junaid Bajwa *et al.*, *op. cit.*

[59] Valentina Bellini *et al.*, *op. cit.*

[60] Barry Solaiman, and I. Glenn Cohen, *op. cit.*

[61] W. Nicholson Price II, «Artificial intelligence in healthcare: applications and legal implications», *The Scitech Lawyer*, 14(1), 2017, p. 10–13.

[62] W. Nicholson Price II, «Artificial intelligence in healthcare: applications and

As a sub-group of ML itself, Deep Learning (DL) uses large numbers of layered structured artificial neural networks (ANN) to interpret the input information, extract features from the data, and provide conclusions in a way similar to what happens for the human brain.[63] ANNs are conceptualized as resembling the functioning of the biological neurons.[64]

The relationship between AI, ML, and DL is represented by Pettit and colleagues in the following diagram (Fig. 1.1).

Fig. 1.1 Representation of concepts: artificial intelligence, machine learning, and deep learning.[65]

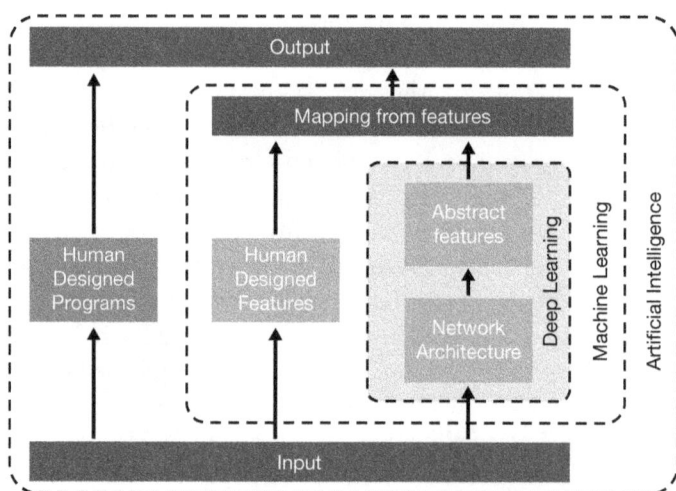

AI systems need large amounts of data both to be trained and to function properly. These data come from various sources, as addressed in the next pages, and may reach the status of big data, which in the absence of a universally agreed definition has been identified as a big «dataset» where the «Log (n*p) is superior or equal to 7» and that may show various characteristics,

---

legal implications», *op. cit.*

[63] Rowland W. Pettit *et al.*, *op. cit.*; Valentina Bellini *et al*, *op. cit.*

[64] Michele Iaselli, «Medicina predittiva e d'iniziativa: l'IA per l'efficientamento della spesa sanitaria», in Ginevra Cerrina Feroni (ed. by), *Le nuove frontiere della medicina*, Il Mulino, 2024, pp. 241-263.

[65] Rowland W Pettit, Robert Fullem, Chao Cheng, Christopher I Amos (2021), «Artificial intelligence, machine learning, and deep learning for clinical outcome prediction», *Emerging Topics in Life Sciences*, 5(6), pp. 729–745.

such as «great variety, high velocity, challenge on veracity and computational methods» etc.[66] Precisely because of the mentioned characteristics, big data is usually impossible to be processed with traditional methods.

The quantity, type, and quality of the data largely influence the functioning of the AI system, and in particular the precision and reliability of its outcomes. As a consequence, the choice of the training dataset is of paramount importance for the successful development of the machine, and in particular to avoid *biases* of the AI. Indeed, in healthcare, an AI system may produce a biased outcome if for instance a biased population sample is chosen as the training dataset, in which a particular subgroup of patients is underrepresented or not at all considered, usually from protected groups. Because the data are collected from the real world, these biases mirror those already present in society and in the provision of healthcare.[67]

Finally, to train AI systems in this field, it is possible to use not only real-world data, i.e., data collected from real patients and real healthcare activities but also synthetic data. Synthetic data is artificially generated information via computer simulations or algorithmic processes that serve as a substitute for actual data from the real world. This type of data provides a viable option when access to real-world data is limited and contributes to enhancing medical and healthcare research by reducing reliance on personal data. Consequently, synthetic data could prove to be an invaluable asset in the advancement of health innovation and pharmaceutical development, addressing the privacy issues that arise from the extensive utilization of data from voluminous clinical trials and electronic health records.[68]

Within the health sector, synthetic data can assist researchers by allowing them to hone and verify methods for specific tasks prior to utilizing actual data, thus accelerating the process and increasing the likelihood of success when applying these methods to real-world data. Additionally, the advent of the Covid-19 pandemic has sparked a surge in interest in synthetic data creation. This includes applications like training AI systems, constructing epidemiological models and digital contact tracing systems, and facilitating data exchanges among healthcare facilities.[69] Additionally, these datasets can contribute to enhancing diversity by replicating the phenotypes of indi-

---

[66] Emilie Baro *et al.*, «Toward a literature-driven definition of big data in health-care», *BioMed Research International*, 1/2015, 2015, pp. 1–9.

[67] Barbara Draghi *et al.*, «Identifying and handling data bias within primary healthcare data using synthetic data generators», *Heliyon*, 10(2), 2024, pp. 1–15.

[68] Theodora Kokosi, and Katie Harron, «Synthetic data in medical research», *BMJ Medicine*, 1(1), 2022, pp. 1–4.

[69] Richard J. Chen *et al.*, «Synthetic data in machine learning for medicine and healthcare», *Nature Biomedical Engineering*, 5, 2021, pp. 493–497.

viduals and conditions that are less commonly represented, expanding the pool of data available for researching uncommon conditions or employing simulations of various processes on algorithmically generated digital twins to assess how outcomes differ. Nevertheless, a significant obstacle to their adoption is determining whether synthetic data will be relied upon for decision-making or if ultimate conclusions must always be drawn from original datasets. Addressing this concern largely depends on the capability to demonstrate that synthetic datasets are valid representations of patient information, thereby being dependable and accurately reflecting reality. With healthcare professionals and decision-makers poised to utilize synthetic data in addressing health-related challenges imminently, establishing an appropriate framework to guarantee its safe application is imperative.

1.4 Concrete applications of AI in healthcare

The first application of AI systems in medicine and for the provision of healthcare can be traced back to the early 1950s.[70] It is at that point in history that «physicians made the first attempts to improve their diagnoses using computer-aided programs».[71]

From that moment onwards, tremendous progress has been made, and nowadays AI is being implemented in almost every area of healthcare, in particular in clinical medicine, biomedical research, public and global health, and healthcare administration.[72] Indeed, as of August 2024, in the USA, the FDA has already approved 950 Artificial Intelligence/Machine Learning enabled medical devices.[73] If compared with the 343 devices approved in 2021,[74] this increase clearly demonstrates the exponential upsurge of AI systems in healthcare.

The engines of this progressive shift toward a more technology-driven

---

[70] Tijs Vandemeulebroucke, *op. cit.*

[71] Chun-Yang Chou, Ding-Yang Hsu, and Chun-Hung Chou, «Predicting the Onset of Diabetes with Machine Learning Methods», *Journal of Personalized Medicine,* 13(3), 2023, pp. 1–18.

[72] Kimberly Badal, Carmen M. Lee, and Laura J. Esserman, «Guiding principles for the responsible development of artificial intelligence tools for healthcare», *Communications Medicine*, 3(47), 2023, pp. 1–6. For a more in-depth analysis of the possible concrete uses of AI for the overall provision of healthcare, see Fei Jiang *et al.*

[73] Intelligence and Machine Learning (AI/ML)-Enabled Medical Devices, FDA website publicly made available list, https://www.fda.gov/medical-devices/software-medical-device-samd/artificial-intelligence-and-machine-learning-aiml-enabled-medical-devices, updated in October 2023.

[74] Kimberly Badal, Carmen M. Lee, and Laura J. Esserman, *op. cit.*

provision of healthcare can be identified, among other things, in some of the serious challenges that the healthcare system had to face over the years, and is still facing, and the increased availability of data, as well as technologies and systems that currently available and that may be implemented.

On the one hand, references can be made, for instance to the Covid-19 health crisis, and to current demographic aging and other lifestyle-related conditions.

In particular, it is well known that Covid-19 forced hospitals and healthcare providers worldwide to rapidly develop and implement new strategies for facing the difficulties caused by the spread of the virus and the high rate of mortality, such as early detecting and diagnosing the infection, providing timely treatments and monitoring it through time and at a distance, projecting cases and mortality, developing drugs and vaccines, and reducing the workload of healthcare workers.[75] The tremendous progress made during the pandemic in terms of technological advancement and adoption of technological solutions in standard and advanced care represented a turning and tipping point, and from that moment onwards, implementation of AI in routine healthcare seems to be a more attainable goal.

As for the second, demographic aging is steadily causing a shift in both the types of health conditions to be treated and the way healthcare is provided. Indeed, the WHO estimated that by 2023, there will be one person in their sixties or older for every six people in the world.[76] The peculiar health-related issues associated with older age are often chronic and comorbid[77] and tend to increase both the demand for healthcare treatments and the overall costs of their provision.[78] For instance, the total global healthcare costs are expected to more than double, rising from 8.4 trillion USD in 2015 to 18.3 trillion USD in 2030. As a consequence, the healthcare system should adjust accordingly to accommodate the specific needs of the elderly, and various AI systems and applications are being developed to be capable of providing a standard of care

---

[75] Among many others, see Raju Vaishya *et al.*, «Artificial intelligence (AI) applications for COVID-19 pandemic», *Diabetes & Metabolic Syndrome*, 14(4), 2020, pp. 337–339; Muzammil Khan *et al.*, «Applications of artificial intelligence in COVID-19 pandemic: a comprehensive review», *Expert Systems with Applications*, 185, 2021, pp. 1–17.

[76] World Health Organization, *Ageing and health – WHO Fact Sheet*, 1st October 2024, https://www.who.int/news-room/fact-sheets/detail/ageing-and-health.

[77] On health-related conditions, see Efraim Jaul, and Barron Jeremy, «Age-related diseases and clinical and public health implications for the 85 Years old and over population», *Frontiers in Public Health*, 5(333), 2017, pp. 1–7.

[78] Adam Bohr, and Kaveh Memarzadeh, «Current healthcare, big data, and machine learning», in Adam Bohr, and Kaveh Memarzadeh (ed. by), *Artificial Intelligence in Healthcare*, Academic Press, 2020, pp. 1–24.

adequate for their conditions, and to adequately predict the probable course of the diseases and their lifestyle more generally, while at the same time managing to reduce the costs and meet the increased demand for care.[79] Reducing costs of care and providing better healthcare thanks to AI are also main goals in the development of preventive medicine for treating diseases and conditions mainly related to lifestyle, such as diabetes and obesity.

On the other hand, there is an ever-increasing availability of personal and health data as part of the more general phenomenon of the datafication of society (and consequently healthcare).[80] Possible sources of the data to be processed for AI purposes include test results, notes of clinicians (also collected in the form of electronic health records, or EHR), medical literature, clinical trials, smart devices (from a simple fitness app on a smartphone to complex medical devices), pharmacy records, insurance claims data,[81] and biobanks, to name the most relevant. In particular, electronic health records had a disruptive impact on healthcare,[82] because they enable the storage of large amounts of health data from each patient throughout his or her care path and life, from different facilities and physicians, as well as in the patient's daily life thanks to the use of smart devices. This information can then be processed in order to better understand the necessities of each individual patient to provide a more personalized healthcare and at the same time to develop care pathways for the entire population based on identified patterns. Thanks to the large amounts of data available and of different types, it has been and still is possible to design and train new algorithms and AI systems to address healthcare needs.

However, the concrete impact of the use of such technologies in the mentioned areas should be carefully evaluated and addressed, both from an ethical and legal perspective, in order to safeguard fundamental rights and consciously guide the development of the technology. We have probably left to science fiction the dystopian future where AI machines completely replace humans, both in everyday activities and in healthcare. Indeed, es-

---

[79] Bingxin Ma *et al.*, «Artificial intelligence in elderly healthcare: a scoping review», *Ageing Research Reviews*, 83, 2023, pp. 1–11; Ching-Hung Lee *et al.*, «Artificial intelligence-enabled digital transformation in elderly healthcare field: scoping review», *Advanced Engineering Informatics*, 55, 2023, pp. 1–20; Srikanta Padhan *et al.*, «Artificial intelligence (AI) and robotics in elderly healthcare: enabling independence and quality of life», *Cureus* 15(8), 2023, pp. 1–4.

[80] In general, on the datafication of healthcare, Minna Ruckenstein, and Natasha Dow Schull, «The datafication of health», *Annual Review of Anthropology*, 46, 2017, pp. 261–278.

[81] W. Nicholson Price II, «Artificial intelligence in healthcare: applications and legal implications», *op. cit.*

[82] Adam Bohr, and Kaveh Memarzadeh, *op. cit.*

pecially in the area of health, it is nowadays apparent that physicians will keep their central and indispensable role in the provision of healthcare to patients. However, the care pathway is, and will continue to be, progressively digitalized by introducing AI systems into the equation. These technologies aim at assisting physicians in making better clinical decisions or in performing a number of tasks (also) on their behalf,[83] researchers in developing new treatments, pharmaceutical or otherwise, and hospital administrators in managing and supervising healthcare facilities.

The adoption of AI technologies in healthcare brings along tremendous concrete advantages that include learning features from a large volume of data, improving the accuracy of clinical decisions, as well as of their own functioning if equipped with self-correcting abilities, providing up-to-date medicinal information to physicians from journals, clinical practices, and many more reducing diagnostic and therapeutic errors, and extracting information to assist in health outcome predictions.[84] This is already being experimented, among other settings, in radiology, pathology, dermatology, ophthalmology, cardiology, emergency triage, and even mental health.[85]

Collectively, all the mentioned fields in which AI can be implemented contribute to bringing contemporary healthcare closer to the ideology of patient-centered care,[86] 4P medicine (predictive, preventive, personalized, and participatory)[87] or, more generally, personalized healthcare. All these «new» conceptualizations of healthcare pivot around some core concepts. First of all, it advocated a shift in healthcare from responding to the outbreak of a disease by providing the appropriate treatment to predicting and preventing it beforehand.

---

[83] Fei Jiang *et al, op. cit.*

[84] *Ibid.*

[85] Tijs Vandemeulebroucke, *op. cit.* For an overview of the possible implementations of AI in healthcare, see also Ahmed Al Kuwaiti *et al.*, «A review of the role of artificial intelligence in healthcare», *Journal of Personalised Medicine*, 13(951), 2023, pp. 1–22.

[86] Traced back to Margaret Gerteis *et al.*, *Through the patient's eyes: understanding and promoting patient-centered care*, Jossey-Bass, 2023.

[87] Leroy Hood *et al.*, «Systems biology and new technologies enable predictive and preventative medicine», *Science*, 306(5696), 2004, pp. 640–664; Andrea D. Weston, and Leroy Hood, «Systems biology, proteomics, and the future of healthcare: toward predictive, preventative, and personalized medicine», *Journal of Proteome Research*, 3(2), 2004, pp. 179–96. Later on, Rob Horne advocated for including a fifth «P», and particularly that of the «psycho-social dimension of care», i.e., the one that cares about patient's informed decision to enable an effective participation. Rob Horne, «The human dimension: putting the person into personalised medicine», *The New Bioethics: A Multidisciplinary Journal of Biotechnology and the Body*, 23(1), 2017, pp. 38–48.

Moreover, the increased availability of data and technologies enables a more participatory provision of healthcare, in which the patient is not confined to receiving and accepting treatments but also plays an active role in the process. This development in healthcare provision is especially enabled, thanks to smart and wearable devices that collect personal health data that can be intertwined with EHR.

Finally, healthcare treatments are evolving in the sense of not providing standardized care pathways for every patient in response to a given disease, but being tailored (as far as possible) to the specific genetic traits of an individual.[88] To do so, access to a massive amount of data, and the computational ability to analyze it, is needed, for which AI represents a valuable tool.[89] This data may include, but is surely not limited to, «personal and family medical history, genetic data, imaging data, information relating to lifestyle and environmental factors».[90]

Even though possible uses of AI in healthcare are expanding continuously, and identifying the most important areas in which AI systems will impact, along the same line as the categorization of the World Health Organisation (WHO),[91] we provide a brief overview of some of the major areas of healthcare and medical practices in which AI can be implemented: diagnosis and clinical care; health research and drug development; and healthcare administration. To the applications identified by the WHO, we added the following: telemedicine; clinical trials; AI in policy development.

1.4.1 AI for diagnosis and clinical care

Medical diagnosis can be broadly defined as the process of evaluating medical conditions or diseases by analyzing symptoms, medical history, and test results, with the goal of determining the causes of a medical problem or condition and providing an effective treatment.

It is first and foremost in this area that AI shows its «transformative im-

---

[88] Ad-Duhaa E. Parekh, *et al.*, «Artificial intelligence (AI) in personalized medicine: AI-generated personalized therapy regimens based on genetic and medical history: short communication», *Annals of Medicine & Surgery*, 85, 2023, pp. 5831–5833.

[89] Kevin B. Johnson *et al.*, «Precision medicine, AI, and the future of personalized healthcare», *Clinical and Translational Science*, 14, 2021, pp. 86–93.

[90] Villaronga *et al.*, «Implementing AI in healthcare: an ethical and legal analysis based on case studies», in Dara Hallinan, Paul de Hert, and Ronald Leenes (ed. by), *Data Protection and Privacy: Data Protection and Artificial Intelligence*, Hart Publishing Ltd., 2021, pp. 187-216.

[91] WHO, *Ethics and governance of artificial intelligence for health: WHO Guidance*, 2021.

pact»[92] and «disruptive potential» for revolutionizing healthcare.[93] In particular, diagnostic imaging represents one of the leading AI applications,[94] especially because of AI's ability to analyze large amounts of data from different datasets and infer correlations among them.[95] These data can be collected from multiple sources and may include medical 2D/3D imaging, biosignals (e.g., ECG, EEG, EMG, and EHR), vital signs (e.g., body temperature, pulse rate, respiration rate, and blood pressure), demographic information, medical history, and laboratory test results.[96] It is precisely by analyzing vast amounts of data from various sources that it is possible to have a more comprehensive and complete picture of the patient's health status and possible development of her health condition.

Concrete advantages of using AI systems in this area of healthcare are primarily related to reducing diagnosis time, costs, and diagnostic errors, catching the outbreak or the presence of a disease earlier than by human intervention, and better-predicting disease development.[97]

Quite recently, various studies have focused on applying ML to predict or diagnose diseases,[98] and NLP to analyze electronic health records with

---

[92] Muhammad Umer Qayyum et al., «Revolutionizing healthcare: the transformative impact of artificial intelligence in medicine», *Bulletin Of Informatics*, 1(2), 2024, pp. 71–83.

[93] Ashish Shiwlani et al., «Revolutionizing healthcare: the impact of artificial intelligence on patient care, diagnosis, and treatment», *Jurihum: Journal Inovasi Dan Humaniora*, 1(5), 2024, pp. 779–790.

[94] Junaid Bajwa et al., *op. cit.*, p. 191.

[95] For more technical and concrete descriptions and analysis of the possible uses of AI in medical diagnosis, see among many others: Vidhya Rekha Umapathy et al., «Perspective of artificial intelligence in disease diagnosis: a review of current and future endeavours in the medical field», *Cureus* vol. 15(9), 2023; Yogesh Kumar et al., «Artificial intelligence in disease diagnosis: a systematic literature review, synthesizing framework and future research agenda», *Journal of Ambient Intelligence and Humanized Computing*, 14(7), 2023; Dóra Göndöcs, and Viktor Dörfler, «AI in medical diagnosis: AI prediction & human judgment», *Artif. Intell. Med.*, 149, 2024; Nafiseh Ghaffar Nia, Erkan Kaplanoglu and Ahad Nasab, «Evaluation of artificial intelligence techniques in disease diagnosis and prediction», *Discover Artificial Intelligence*, 3(5), 2023.

[96] Mugahed A. Al-Antari, «Artificial intelligence for medical diagnostics—existing and future AI technology!», *Diagnostics*, 13(688), 2023.

[97] Shuroug A. Alowais et al., *op. cit.*

[98] Marwa Elgenedy et al., «An MRI-based deep learning approach for accurate detection of Alzheimer's disease», *Alexandria Engineering Journal*, 63, 2023, pp. 211–221; Yogesh H. Bhosale, and K Sridhar Patnaik, «PulDi-COVID: Chronic obstructive pulmonary (lung) diseases with COVID-19 classification using ensemble deep convolutional neural network from chest X-ray images to minimize severity and mortality rates», *Biomedical Signal Processing and Control*, 81, 2023.

the aim of extracting information to be used in disease prevention and diagnosis.[99]

In particular, while there is still a long way to go to fully and routinely implement AI in diagnosis, tremendous successes have been made in radiology and medical imaging. For instance, a study from 2021 showed that AI could diagnose breast cancer with mass more precisely (90%) than radiologists (75%) and also could better detect early breast cancer.[100] Another study showed that pneumonia could be detected from chest radiography by AI with better sensitivity and specificity (96% and 64%) than radiologists (50% and 73%, respectively),[101] and it is also suggested that it could detect infections (COVID-19, for instance), using blood specimens or images.[102] Similarly, AI has been tested and used for predicting various pathologies for which not only precision, but also detection is vital, such as liver disease, heart disease, Alzheimer's disease, and various types of cancer.[103]

Diagnosis and prediction are also breeding grounds for black-box medicine, such as for identifying skin cancer,[104] predicting the probability of

---

[99] Chengtai Li *et al.*, «Natural language processing applications for computer-aided diagnosis in oncology», *Diagnostics*, 13(286), 2023; Mahmud Omar *et al.*, «Utilizing natural language processing and large language models in the diagnosis and prediction of infectious diseases: a systematic review», *American Journal of Infection Control*, 52(9), 2024, pp. 992–1001.

[100] Kim Hyo-Eun *et al.*, «Changes in cancer detection and false-positive recall in mammography using artificial intelligence: a retrospective, multireader study», *The Lancet. Digital health*, 2(3), 2020, pp. 138–148.

[101] Judith Becker *et al.*, «Artificial intelligence-based detection of pneumonia in chest radiographs», *Diagnostics*, 12(1465), 2022.

[102] Maad M. Mijwil, and Karan Aggarwal, «A diagnostic testing for people with appendicitis using machine learning techniques», *Multimedia Tools and Applications*, 81(5), 2022, pp. 7011–7023.

[103] Rohit Bharti *et al.*, «Prediction of Heart Disease Using a Combination of Machine Learning and Deep Learning», *Computational Intelligence and Neuroscience*, 2021(8387680), 2021; Pradyumna Byappanahalli Suresha *et al.*, «A deep learning approach for classifying nonalcoholic steatohepatitis patients from nonalcoholic fatty liver disease patients using electronic medical records», in Arash Shaban-Nejad, Martin Michalowski, and David L. Buckeridge (ed. by), *Explainable AI in Healthcare and Medicine*, Springer, 2021, pp. 107-113; Nitika Goenka, and Shamik Tiwari, «Deep learning for Alzheimer prediction using brain biomarkers», *Artificial Intelligence Review*, 54(7), 2021, pp. 4827-4871.

[104] Andre Esteva *et al.*, «Dermatologist-level classification of skin cancer with deep neural networks», *Nature*, 542(7639), 2017, pp. 115–118.

trauma victims of hemorrhaging,[105] and suggesting off-label uses of autho-rized drugs.[106]

Moreover, AI can also be coupled with developments in genomic medi-cine to identify genetic markers linked to genetic diseases and thus predict and prevent outbreaks of disease threats.[107] The ability of AI systems to ana-lyze a large quantity of genetic data and infer genetic information from them contributes to the development of healthcare toward a more personalized approach that brings the patient at the center of the care pathway, identified based on the unique genetic characteristics, clinical history, prognostic in-formation, and treatment response of the person.[108]

In particular, the aim of personalized medicine has been defined as «to improve stratification and timing of healthcare by utilising biological in-formation and biomarkers on the level of molecular disease pathways»,[109] even though no complete personalisation of any treatment may or should be achieved.[110]

Steps further in the tailoring of healthcare to individuals have also been taken, thanks to smart and wearable devices that are able to collect large amounts of everyday health data from patients, even outside the healthcare facilities, and are linked to one another thanks to the rise of the Internet of Things (IoT).[111] Wearable devices, in particular, are all those devices that can be worn by the patient, in various forms such as wristbands or glasses[112] and are therefore able to constantly monitor the fitness level or other param-eters of the owner.[113] Not only can these devices help constantly monitor the health parameters and overall data of a single patient but also they can

---

[105] Nehemiah T. Liu *et al.*, «Development and validation of a machine learning al-gorithm and hybrid system to predict the need for life-saving interventions in trauma patients», *Medical & Biological Engineering & Computing*, 52(2), 2014, pp. 193–203.

[106] W. Nicholson Price II, «Artificial intelligence in healthcare: applications and legal implications», *op. cit.*, p.1.

[107] Shuroug A. Alowais *et al.*, *op. cit.*

[108] Adam Bohr, and Kaveh Memarzadeh, *op. cit.*

[109] Jochen Vollmann, Verena Sandow, and Jan Schildmann, *The ethics of person-alised medicine: Critical perspectives*, Routledge, 2015.

[110] Therese Feiler *et al.*, «Personalised medicine: the promise, the hype and the pit-falls», *The New Bioethics: A Multidisciplinary Journal of Biotechnology and the Body*, vol. 23(1), 2017, pp. 1–12.

[111] Arash Aframian *et al.*, «Medical devices and artificial intelligence», in Adam Bohr, and Kaveh Memarzadeh (ed. by), *Artificial Intelligence in Healthcare*, Academic Press, 2020, pp. 163-177

[112] Farida Sabry *et al.*, «Machine learning for healthcare wearable devices: the big picture», *Journal of Healthcare Engineering*, 2022(4653923), 2022.

[113] Maurizio Rizzetto, «Wearables e IoT per il monitoraggio costante del paziente a distanza: opportunità e limiti», in Ginevra Cerrina Feroni (ed. by), *Le nuove frontie-re della medicina*, Il Mulino, 2024, pp. 153-161.

also be used for early diagnosis or increasing adherence with already decided clinical pathways.[114] By interacting with the device, patients can feel more engaged in the medical decisions, and this can overall contribute to increasing the quality of the provided healthcare. Moreover, these devices collect huge quantities of health data, especially on the interaction between humans and the AI technology,[115] which can then be further analyzed and processed for various purposes, including developing new AI systems and devices, provided that the legal requirements for data protection are complied with.

Finally, in the context of AI used to concretely provide (better) healthcare, one of the most promising applications are also virtual conversational agents, or chatbots, that can dialogue with and inform the patient in a more personalized way, according also to his knowledge level and thus assisting him primarily, and asking him specific questions, the answers to which are an invaluable source of information for the clinician for better assisting the patient.[116]

An example of a wearable device for disease diagnosis was the one developed by Hssayeni et al. to detect early signs of Parkinson's disease from the physical movement of patients, using accelerometers and gyroscopes,[117] and the data collected by them, or the Alzheimer's detecting device, that used accelerometers to identify potentially problematic walking patterns.[118]

1.4.2 AI for telemedicine

As per the definition provided for by the World Health Organization, telemedicine is the use of information and communication technologies to provide healthcare services when distance is a critical factor.[119] Various healthcare services may be provided within the broader category of telemedicine, such as diagnosing, treating, preventing diseases and injuries, conducting

---

[114] Darius Nahavandi *et al.*, «Application of artificial intelligence in wearable devices: Opportunities and challenges», *Computer Methods and Programs in Biomedicine*, 213, 2022.

[115] Farida Sabry *et al.*, *op. cit.*

[116] Villaronga *et al.*, *op. cit.*

[117] Murtadha. D. Hssayeni *et al.*, «Wearable Sensors for Estimation of Parkinsonian Tremor Severity during Free Body Movements», *Sensors*, 19(4215), 2019, pp. 1-17.

[118] Ramachandran Varatharajan *et al.*, «Wearable sensor devices for early detection of Alzheimer disease using dynamic time warping algorithm», *Cluster Computing*, 21(1), 2018, pp. 681–690.

[119] WHO Group Consultation on Health Telematics, *A health telematics policy in support of WHO's Health for-All strategy for global health development: Report of the WHO group consultation on health telematics*, 1997.

research, and providing continuing education,[120] and a wide number of tools may be deployed to this end, among which AI plays an important role.

Telemedicine strongly developed and became widely used during the Covid pandemic from 2019, when possibilities of reaching healthcare facilities were limited and private houses had to be somehow transformed into the new points of care. As a consequence, patients' needs could not be met without the adoption of innovative methods and the implementation of AI systems to constantly monitor them, dialogue, and provide the required healthcare service. Moreover, these systems were also used to enable communications between hospitals and thus cooperation among healthcare professionals, frequently for collaborating on a given clinical case.

More precisely, nowadays four trends of AI in telemedicine can be identified: (1) monitoring of patients, (2) using information technology in healthcare, (3) use of intelligent assistance and diagnosis, and (4) collaborative information analysis.[121] All the mentioned healthcare goals may be reached, thanks to the use of «(a) mobile health tools, such as wearables, devices, and smartphone/tablet/laptop applications, collectively described as mobile health (mHealth) tools (b) remote patient monitoring platforms used to manage chronic care diseases, e.g., diabetes, heart failure (HF), hypertension (HTN), obesity, (c) audio and visual connections that enable real-time clinical encounters between a clinician and their patients, (d) image and data collection, storage and forwarding for subsequent asynchronous evaluation, interpretation, and consultation, and (e) virtual updating of the electronic health record (HER) system to allow for virtual check-in's and secure communication between clinicians and their patients, usually via messaging».[122]

Introducing AI in telemedicine has fundamental advantages, which include increasing access to healthcare (enabling the possibility of reaching a wider range of patients), improving communication between healthcare providers and improving efficiency of the healthcare system as a whole, by reducing hospitalization and decreasing costs.[123] At the same time, some of

---

[120] Sachin Sharma, Raj Rawal, and Dharmesh Shah, «Addressing the challenges of AI-based telemedicine: best practices and lessons learned», *Journal of Education and Health Promotion*, 12(338), 2023.

[121] Danica Munson Pacis, Edwin Jr. Dela Cruz Subido, and Nilo T. Bugtai, «Trends in telemedicine utilizing artificial intelligence», *AIP Conference Proceedings*, 13(1933), 2018.

[122] Efstathia Andrikopoulou, «The rise of AI in telehealth», in Andrew Freeman, and Ami Bhatt, *Emerging Practices in Telehealth - Best Practices in a Rapidly Changing Field*, Academic Press, 2023, pp. 183-207.

[123] Ayesha Amjad, Piotr Kordel, and Gabriela Fernandes, «A review on innovation in healthcare sector (telehealth) through artificial intelligence», *Sustainability*, 15(8), 2023.

the risks and possible disadvantages associated with this new way of delivering healthcare are related to the challenges of the digital divide and digital literacy, costs of the development of such technologies, and the need to adequately process data and protect them, especially from cyberattacks.

1.4.3 AI for health research and drug development

Notwithstanding the prevalent conservative approach of the pharmaceutical industry,[124] preserved and incentivized also by regulatory choices,[125] drug discovery and development can benefit to a large extent from including AI systems in the process.[126]

Possible applications of AI in the drug lifecycle include drug discovery and design, predicting drug properties, identifying targets, discovering hits, optimizing leads, predicting ADMET outcomes, clinical trial design, and product management overall in the subsequent phases,[127] even though unfortunately it is sometimes relegated to marginal and highly specialized tasks notwithstanding its strong potential.[128]

In the preclinical stages of drug development, AI systems are primarily used for finding and proposing new chemical compounds,[129] a task that requires expert knowledge of the biological functioning of the human body and the ability to analyze large amounts of technical data, and infer correlations among them.[130] The challenges faced in drug design include the time-intensive and expensive nature of the process, which demands substantial resources. It is believed that launching a new pharmaceutical prod-

---

[124] Stefano Colombo, «Applications of artificial intelligence in drug delivery and pharmaceutical development», in Adam Bohr, and Kaveh Memarzadeh (ed. by), *Artificial Intelligence in Healthcare*, Academic Press, 2020, pp. 85–116.

[125] Veer Patel, and Manan Saha, *Artificial Intelligence and machine learning in drug discovery and development*, MedNexus, 2022.

[126] Óscar Álvarez-Machancoses, and Juan Luis Fernández-Martínez, «Using artificial intelligence methods to speed up drug discovery», *Expert Opinion on Drug Discovery*, 14(8), 2019, pp. 769–777; Marwin Segler *et al.*, «Generating focused molecule libraries for drug discovery with recurrent neural networks», *ACS Central Science*, 4(1), 2018, pp. 120–131.

[127] Linda Nene *et al.*, «Evolution of drug development and regulatory affairs: the demonstrated power of artificial intelligence», *Clinical Therapeutics*, 46(8), 2024, pp. 6–14.

[128] Stefano Colombo, *op. cit.*

[129] Rohan Gupta *et al.*, «Artificial intelligence to deep learning: machine intelligence approach for drug discovery», *Molecular Diversity*, 25(3), 2021, pp. 1315–1360.

[130] Bowen Lou, Lynn Wu, «AI on drugs: can artificial intelligence accelerate drug development? Evidence from a large-scale examination of bio-pharma firms», *MIS Quarterly*, 45(2021), 2021, pp. 1451–1482.

uct can cost around $2.5 billion, with the chance of moving a drug from discovery to clinical trials at about 35%.[131] Moreover, the probability of advancing from Phase 1 clinical trials to obtaining regulatory approval is only between 9% and 14%, typically requiring between 12 and 15 years.[132] Artificial intelligence and machine learning could be the answer to these obstacles, offering the necessary assistance for researchers.[133] It has been estimated that to propose a single molecule about 10,000 chemical or biological compounds are screened.[134] In this, machines obtain results with a higher degree of novelty than human experts using traditional methods.[135] Indeed, it is in this stage of the drug development that AI is most beneficial, considering for instance the possibility for supervised algorithms to infer conclusions that are not directly instructed by humans.[136] AI can thus propose hypotheses for new drug compounds with a desired beneficial effect that could not be otherwise envisioned by experts, and could help address the most complex goal in drug development, i.e., finding new drugs for novel conditions and therapies. This means that AI and machine learning will aid scientists in pinpointing the best drug molecules that present minimal risk, within a regulatory framework that enforces strict safety and efficacy testing for new drugs.

Other possible uses of AI systems in this field include predicting and estimating tasks, such as the probability of success of a new drug. To this end, AI systems could possibly process real-world data[137] or predict various drug functions and reducing the need to include real-world patients in clinical trials.

As a consequence, AI can help address some of the major concerns related to drug discovery, which include its costs in terms of finances and time,[138]

---

[131] *Unlocking the potential of AI in Drug Discovery*, Wellcome Trust, 2023, https://wellcome.org/reports/unlocking-potential-ai-drug-discovery

[132] *Ibid.*

[133] Włodzisław Duch *et al.*, «Artificial intelligence approaches for rational drug design and discovery», *Current Pharmaceutical Design*, 13(14), 2007, pp. 1497–508.

[134] Dnyaneshwar Kalyane *et al.*, «Artificial intelligence in the pharmaceutical sector: current scene and future prospect», in *The Future of Pharmaceutical Product Development and Research*, 2020, pp. 73–107.

[135] Anne Trafton, *Artificial Intelligence Yields New Antibiotic*, MIT News, 2020.

[136] Bowen Lou, Lynn Wu, *op. cit.*

[137] Simon Dagenais *et al.*, «Use of real-world evidence to drive drug development strategy and inform clinical trial design», *Clinical Pharmacology and Therapeutics*, 111(1), 2022, pp. 77–89.

[138] Dnyaneshwar Kalyane *et al.*, *op. cit.*

and at times poor quality of the results.[139] In particular, AI can perform human tasks to be completed in this process 100 times faster than humans.[140]

At the same time, similar results can also be reached by asking AI to propose new uses of existing drugs for treating medical conditions not envisioned at the time of the drug development or approval (drug repurposing),[141] an approach that helps eliminate the issues related to drug novelty. Also asking AI to develop a «drug molecule that has an inherent ability to interact with multiple targets or pathways within the domain of disease-related molecular networking», thus adopting the approach of polypharmacology can be extremely beneficial in terms of costs and time-saving.[142]

Indeed, AI role in research and development in assisting in the determination of new therapeutic targets or finding new uses for existing medications is facilitated by its capability to sift through vast and intricate datasets, including genomic, proteomic, transcriptomic, and clinical data, to generate new hypotheses and insights. AI aids not just in forming these hypotheses but also in their validation and prioritization, and in planning and executing experiments. Furthermore, predictive models known as safety risk score models are deployed to foresee the possible toxicity and side effects of drug candidates, and AI is instrumental in enhancing these models. By drawing on diverse data sources, such as preclinical studies, clinical trials, pharmacovigilance reports, and real-world evidence, AI contributes to identifying and measuring the safety indicators and risk elements linked to drugs.

Lastly, AI methodologies can also be implemented in the determination of dosages, that is to say, calculating appropriate drug amounts and scheduling for patients. AI has the capacity to personalize and fine-tune dosage plans relative to individual patient factors like age, body mass, genetic profile, biomarkers, existing health conditions, and concurrent medication use. Moreover, AI assistance is instrumental in tracking and revising the dosage regime in response to how the patient reacts, complies with the treatment plan, and reports outcomes.

It's crucial to note that predictive modeling has long been integral to research and development efforts, with animal testing being one of the most ubiquitous forms. Although animal testing furnishes invaluable insights into drug safety and potential effectiveness, it does not guarantee that results in

---

[139] Linda Nene *et al.*, *op. cit.*

[140] William Smith, *How Atomwise Uses Artificial Intelligence for Drug Discovery*, Medium, 2020.

[141] Francesco Napolitano *et al.*, «Drug repositioning: a machine-learning approach through data integration», *Journal of Cheminformatics*, 5(30), 2013.

[142] Dnyaneshwar Kalyane *et al.*, *op. cit.*

animals will translate directly to humans. Similarly, predictive AI forecasts possible outcomes based on available data. Presently, AI primarily serves as an ancillary instrument in R&D, complementing but not replacing conventional techniques. As we look ahead, AI might take precedence as the primary prognostic tool in these fields.

As these burgeoning technologies are streamlining the R&D process, enhancing cost-effectiveness, and speeding up the processes, it is incumbent upon regulators to consider these advantages while crafting laws. The obstacle here is securing high-quality data, particularly where the culture of open data exchange in healthcare and pharmaceutical sectors is still relatively rare.

1.4.4 AI for healthcare administration

AI can be integrated into hospitals in various ways. In the previous pages, AI systems were presented as generally used for providing current healthcare (for diagnosis and clinical care) or developing new care pathways (for clinical care).

However, AI can also be implemented to better *administer* the way in which healthcare is provided, and it is frequently advocated that more digitalized healthcare facilities and management processes will reduce costs, increase efficiencies and overall increase quality of care,[143] as it is said to have the potential to «revolutionise the landscape of hospital management».[144]

While any attempt at classifying the possible implementation of AI in hospital management would inevitably be a simplification of the complexity of the managerial structure of healthcare facilities and workflow, Malik and Solaiman in particular identified that AI in this field can be used for predictive maintenance, bed management, and staff rota planning.[145] These tasks are usually not directly related to the patient, or do not modify the physician-patient relationship directly, but aim at overall providing better healthcare.

---

[143] Xinghua Gao, and Pardis Pishdad-Bozorgi, «BIM-enabled facilities operation and maintenance: a review», *Advanced Engineering Informatics*, 39, 2019, pp. 227–247; Krystyna Araszkiewicz, «Digital technologies in facility management—the state of Practice and research challenges», *Procedia Engineering*, 196, 2017, pp. 1034–1042.

[144] Shefali Bhagat, and Deepika Kanyal, «Navigating the future: the transformative impact of artificial intelligence on hospital management—a comprehensive review», *Cureus*, 16(2), 2024.

[145] Abeer Malik, and Barry Solaiman, «AI in hospital administration and management: ethical and legal implications», in Barry Solaiman, and Glenn Cohen, *Research Handbook on Health, AI and the Law*, Edward Elgar Publishing, 2024, pp. 20-40.

More specifically, predictive maintenance aims at identifying possible causes for unexpected failures of medical equipment, and therefore at reducing and preventing them. This goal may be reached by analyzing «substantial amounts of time-sensitive data from individual in-hospital equipment separately» thanks to AI algorithms and techniques,[146] and that can be turned into «remote monitoring, automated control of facility systems, immediate awareness of emergent conditions, predictive maintenance instead of break fixes, and countless labor-saving opportunities that reduce staffing».[147] However, predictive maintenance should be coupled with ordinary maintenance, scheduled at regular intervals to take corrective actions. High quality and efficiency of the equipment are strictly related to providing a high standard of care and reducing costs, as it has been estimated that maintenance represents 70%–80% of the costs of a facility asset[148] and that implementing predictive maintenance could reduce up to 98% of the total maintenance costs.[149]

Moreover, AI technologies can be used for more efficient and effective bed management,[150] and in general patient flow, especially in emergency settings, where delays in admission represent a «well-defined risk factor».[151] As for other areas of healthcare, both aspects of hospitals and healthcare facilities' management were particularly cause for concern during the outbreak of Covid-19, and are generally associated with reduced costs and shorter length of stay,[152] as well as higher quality of care for patients, and operational efficiency. In 2017, a hospital in Aosta had already tested a digitalized bed management system, with great results in terms of decreasing bed occupancy and outlier days, and Emergency Department admissions.[153]

Finally, AI can be implemented in hospital management of staff rota

---

[146] *Ibid.*

[147] Anthony D. Scaife, «Improve predictive maintenance through the application of artificial intelligence: a systematic review», *Advance*, 21, 2024.

[148] Dania K. Abideen *et al.*, «A systematic review of the extent to which BIM is integrated into operation and maintenance», *Sustainability*, 14 (8692), 2022; Naseer Muhammad Khan *et al.*, «Development of predictive models for determination of the extent of damage in granite caused by thermal treatment and cooling conditions using artificial intelligence», *Mathematics*, 10 (16), 2022.

[149] Douglas S. Thomas, «The costs and benefits of advanced maintenance», in *Advanced Manufacturing Series*, 2018, pp. 1–45.

[150] Abeer Malik, and Barry Solaiman, *op. cit.*

[151] Roberto Novati *et al.*, «Effectiveness of an hospital bed management model: results of four years of follow-up», *Annali di igiene: medicina preventiva e di comunità*, 29(3), 2017, pp. 189–196.

[152] Qing Huang *et al.*, «The impact of delays to admission from the emergency department on inpatient outcomes», *BMC Emerging Medicine*, 10(16), 2010.

[153] Roberto Novati *et al.*, *op. cit.*

planning, which is the «schedule or list that assigns tasks or responsibilities to individuals in a rotating or sequential order»,[154] that can be perfectionated using AI technologies.

1.4.5 AI for clinical trials

Delving into a different aspect of the drug development process, AI technologies are making strides in clinical trials by aiding in the identification and recruitment of suitable patients, as well as handling the vast amounts of data generated.[155] Additionally, AI has the potential to refine the design of clinical trials by predicting patient outcomes and the likelihood of a trial's success, which could lead to fewer participants required and shorter trial durations. This has the potential to expedite the clinical development process, increasing the chances of trial success and regulatory approvals. Consequently, the benefits of AI in clinical trials are substantial, albeit they come with unique regulatory hurdles and risks compared to earlier applications. Therefore, establishing validation standards for the algorithms used in the process and implementing safeguards to prevent the reinforcement or emergence of biases is crucial.

Indeed, one concerning potential issue in this field is the use of an algorithm designed to determine and choose participants for a clinical trial. If such an algorithm exhibits bias, it could prevent marginalized communities from obtaining healthcare, exacerbating existing disparities and infringing on individuals' health rights. Additionally, the integrity of the dataset is vital, and implementing transparency protocols might be required to detect any possible biases. Notably, under the European AI Act, an AI system must be deemed high-risk if it entails profiling individuals.

1.4.5 AI in policymaking

The integration of AI and machine learning in public health policy formulation is crucial due to its capacity to process extensive healthcare data and generate valuable insights.[156] By examining complex datasets, machine learning algorithms can detect patterns, identify trends, and unveil potential risk fac-

---

[154] Abeer Malik, and Barry Solaiman, *op. cit.*

[155] Hitesh Chopra *et al.*, «Revolutionizing clinical trials: the role of AI in accelerating medical breakthroughs», *International Journal of Surgery*, 109(12), 2023, pp. 4211–4220.

[156] Maryam Ramezani *et al.*, «The application of artificial intelligence in health policy: a scoping review», *BMC Health Services Research*, 23(1416), 2023.

tors related to various health conditions and diseases. Health authorities can leverage LLMs to track references to symptoms, travel behaviors, and public attitudes, facilitating the early identification of potential outbreaks. This prompt data allows officials to quickly deploy response strategies, including targeted interventions and resource distribution, to control and reduce the effects of new health threats. Additionally, machine learning supports predictive modeling, allowing policymakers to forecast healthcare needs and allocate resources efficiently, thereby enhancing healthcare delivery. It also helps identify vulnerable populations and health inequalities, aiding in the creation of focused interventions to tackle specific healthcare problems. AI can also enhance vaccination distribution and uptake by forecasting demand and pinpointing high-risk groups. Finally, AI can help predict and prevent noncommunicable diseases by examining risk factors, genetics, and lifestyle trends.

However, the efficacy of machine learning recommendations and deep learning insights in public policy hinges not only on the technology's robustness but also on the quality of the input data, given that AI systems are known to suffer from biases in their inputs, processing, and outputs. These social biases can pose serious risks related to AI deployment, as they can lead to unequal outcomes among patient populations and protected demographic groups. Therefore, it is essential to scrutinize how algorithms are designed as they often perpetuate patterns of inequality. Furthermore, implementing robust data privacy and security protocols, including encryption, access controls, and adherence to data protection laws, is crucial for protecting sensitive health information in AI applications within public health.[157]

1.5 Conclusions

Notwithstanding the mentioned difficulties in reaching an agreement over the constitutive elements of AI and its (legal) definition, AI is being implemented in various areas of healthcare, from the development of new care pathways to the concrete management of the ways in which healthcare is provided. In this sense, the advantages related to its implementation in everyday clinical and care practice are enormous, in terms of ameliorating quality of care, increasing the level of personalization of the care pathway

---

[157] Pan American Health Organization, *Q&A on artificial intelligence for supporting public health—Reference tool to support the exchange of information and promote open conversations and debates*, 2024.

while at the same time decreasing its time and costs, and generally reducing staff workload.

However, alongside these major upsides, there are also disadvantages that should be addressed. Some of them have been mentioned already in this chapter, such as algorithm biases, the need for high quality and quantity of data, and some others are strictly related to the environment in which the machines are implemented and used, as costs of development and maintenance and the level of digital literacy of those who shall use and interact with AI systems. Moreover, there are complex policy, ethical, and legal issues that arise from the described concrete implementations of AI and also from the mentioned concrete disadvantages that should be addressed in order to ensure that the use of AI systems in healthcare complies with sound ethical and legal standards, and at the same time addresses societal needs.

After this first overview of the fields and the issues at stake, the next Chapter will be devoted to dive into the concrete aspects of AI in contemporary society.

PART II
KEY ETHICAL, LEGAL, AND POLICY CHALLENGES IN REGULATING ARTIFICIAL INTELLIGENCE IN HEALTHCARE

2 Legal Theory Part. Addressing the Ethical and Legal Challenges of the Design, Development, and Deployment of AI in Healthcare

2.1 Introduction

Alongside the wide variety of policy issues, the expansion of the adoption of AI systems for the ordinary provision of healthcare and their role «as [drivers] of structural change rather than just [tools]»[1] brings along ethical and legal concerns. Indeed, on the one hand, the former is related to the need to identify ethical principles that could guide stakeholders toward an ethically sound development and use of such technology, both in general and for the specific context of healthcare (frequently referred to as AI Ethics[2]), while the latter focus on providing enforceable rules to adequately safeguard fundamental rights and interests worthy of protection in the field.

The underlying idea behind both sets of rules, ethically and normatively characterized, is to guide the development and subsequent deployment of the technology throughout its entire lifecycle so that it is worthy of the trust of all the stakeholders, actors, and people involved at various levels and stages of its gradual roll-out in the healthcare system and ultimately deployed for the benefit of mankind.

Indeed, as it frequently happens in the field of new technologies, also AI has been mostly developed in a *legal* vacuum, in the sense that up to very recently, there were almost no legally enforceable norms to be complied with either in the design of such technology or for its concrete deployment, especially within the healthcare system. However, as already mentioned, a regulatory framework is needed in order to prevent or address potential violations of fundamental rights or endangering *constitutional* values protected under each country's jurisdiction, as well as at the international level.

The theoretical pathway to a regulation of this kind (i.e., the pathway from «AI to law»), especially if to be enacted for the healthcare sector, has

---

[1] Melanie Smallman, «Why we need to consider the ethics of AI in healthcare at different scales», *Science and Engineering Ethics*, 28(63), 2022, pp. 1–17.

[2] Paula Boddington, *AI ethics—A textbook*, Springer, 2023.

been conceptualized as a «four-stages process.»[3] (1) At the beginning, almost every new technology is being developed and used with no legal rules specifically applicable to it, bringing along great advantages but also raising concerns for potential (unforeseeable) consequences, which are frequently unique to the concrete technology under consideration, and (2) attempts are made to regulate the matter by interpreting existing norms and trying to adapt them to the new realm. To address any issues that cannot be solved by analogy or by extensively interpreting the available existing framework, (3) principle-based[4] soft law instruments or ethical documents are published, in the form of guidelines, ethics codes, policies, or otherwise, and in turn (4) these will eventually become the basis for the development of new norms tailored to the specific matter and technology under consideration.[5]

In some jurisdictions, when AI is considered, this process has nowadays reached the final stage; thus, some instances of regulations are currently available around the globe, such as the European AI Act, most of which are explicitly based on several of the ethical guidelines discussed in the following pages. However, these newly enacted norms are still in their infancy and are usually not tailored to the specific field of healthcare, to which only some norms are usually dedicated, at most. Therefore, it is worth providing an overview of the ethical principles surrounding the implementation of AI in healthcare and of the theoretical legal issues that arise from its design, development, and use in order to understand which rails should be followed by future legal developments and for the interpretation of existing norms in the field under consideration.

Indeed, on the one hand, ethical principles should be considered not only when structuring new regulations but also when interpreting already enacted norms, as well as in the concrete deployment of the regulated technology and its future developments. While norms instruct stakeholders on *what could be done*, ethical principles guide the (consequential) decision about *what should be done*, based on the assumption that not every compliant and legitimate action (or AI system) according to the law is at the same time «ethically justifiable or socially acceptable.»[6] Providing an overview of the relevant ethical concerns surrounding the development of AI systems and their use,

---

[3] Barry Solaiman, «From 'AI to law' in healthcare: the proliferation of global guidelines in a void of legal uncertainty», *Medicine and Law*, 42(2), 2023, pp. 391-406.
[4] Jessica Morley *et al.*, «Ethics as a service: a pragmatic operationalisation of AI ethics», *Minds and Machines*, 31, 2021, pp. 239-256.
[5] Barry Solaiman, *op. cit.*
[6] Jessica Morley *et al.*, *op. cit.*

especially in healthcare, remains relevant for evaluating both existing and future norms and stakeholders' behaviors.

On the other hand, the enacted regulations on the matter (discussed in the following chapters) do not address any and all of the legal issues to be tackled, nor do regulations in different jurisdictions provide the same solutions. It is thus interesting to provide a general overview of the issues that should theoretically be addressed by a regulation on the matter and then leave the discussion of the concrete approaches already adopted for the following chapters. The legal issues discussed in the following pages will be both those generally relevant for regulating AI and those specific for addressing AI in healthcare.

The principles that contribute to shaping AI Ethics vary according to the concrete application of AI under consideration. Indeed, the ethical principles that govern the use of AI systems for the provision of healthcare directly to patients, such as, among many others, decision-support systems, may differ in some regards from those to consider when using AI systems in the context of drug development. Differences are mainly due to the diversity of interests and rights that are safeguarded in each context and given situation. In the following pages, we provide a general overview of AI Ethics and highlight specificities when relevant. In particular, for the purposes of this chapter, we believe that the concrete application of AI in healthcare may be grouped into applications with *direct* or *indirect* consequences on patients. With reference to the applications discussed in Chapter 1, the first group (application with direct consequences on patients) would include most of the AI systems used for healthcare administration, diagnosis, and clinical care, while in the second pertains to health research and drug development, as well as the remaining AI applications for healthcare administration, diagnosis, and clinical care.

2.2 Ethical challenges

The envisioned disruptive impact of AI in society and more specifically in healthcare called for the establishment of ethical principles already in the infancy of the development of the technology. Indeed, the ethical debate on the matter can be traced back to 1960 but exploded in recent years, with the AI Ethics Guidelines Global Inventory that already in April 2020 listed 167 guidelines around the globe.[7]

---

[7] See for instance Wiener, Norbert, «Some moral and technical consequences of automation», *Science,* 131(1340), 1960, pp. 1355-1358; Arthur Samuel, «Some moral and technical consequences of automation—a refutation», *Science,* 132(3429), 1960,

This proliferation of documents addressing the ethical issues of AI, in general and not only when specific fields of application are considered (of which healthcare is one of the most prominent and frequently scrutinized) raised concerns among scholars on the possibility of this proliferation of principles to become «overwhelming and confusing» for those who need to apply them.[8] On a more concrete level, a non-coherent ethical framework that could result from the vast number of guidelines and documents available for stakeholders could facilitate what has been defined as a «market for principles» or «ethics shopping» according to which stakeholders could take advantage of the amount and diversity of ethical principles available to choose those that are more favorable to their specific situation instead of selecting and complying with the ethical principles that best address the issues raised by it.[9] However, opportunistically selecting ethical principles directly contradicts their ontological purpose.

On the matter, Floridi and Cowls provide a positive perspective by highlighting that six high-profile initiatives in the ethical domain[10] show «coherence and overlap,» especially when the specific principles thereby proposed are compared to the four traditional bioethical principles,[11] namely beneficence, non-maleficence, autonomy, and justice firstly developed by Beauchamp and Childress.[12] In the specific context of AI, as it will be further discussed in the following pages, they suggest that explicability should be added to the equation as a fifth basic principle, comprising both intelligibility and accountability.[13]

---

pp. 741-742. On this, Luciano Floridi, *Etica dell'intelligenza artificiale – Sviluppi, opportunità, sfide*, Scienza e idee, 2022.

[8] Luciano Floridi, and Josh Cowls, *op. cit.*

[9] Luciano Floridi, «Translating principles into practices of digital ethics: five risks of being unethical» *Philosophy & Technology*, 32, 2019, pp. 185-193.

[10] The chosen initiatives were the Asilomar AI Principles, the Montreal Declaration for Responsible AI, the second version of the Ethically Aligned Design: A Vision for Prioritizing Human Well-being with Autonomous and Intelligent Systems, the Statement on Artificial Intelligence, Robotics and 'Autonomous' Systems, the «Five overarching principles for an AI code' offered in UK House of Lords Artificial Intelligence Committee's report, and the Tenets of the Partnership on AI.

[11] Luciano Floridi, and Josh Cowls, *op. cit.*

[12] Tom Beauchamp, and James Childress, *Principles of biomedical ethics*, Oxford University Press, New York, 2013.

[13] Luciano Floridi, and Josh Cowls, *op. cit.*; Similarly, Tugba Akinci D'Antonoli *et al.*, «Ethical considerations for artificial intelligence implementation in radiology», *European Society of Radiology*, 2019, pp. 1–13; Tugba Akinci D'Antonoli, «Ethical considerations for artificial intelligence: an overview of the current radiology landscape. Diagnostic and interventional radiology», *Diagnostic and Interventional Radiology*, 26, 2020, pp. 504-511.

Overall, the aim of most of the ethical guidelines and documents is to guide technological development in a manner that foster a trustworthy a trustworthy and human-centered AI. On the one hand, fostering people's trust in the use and results of AI is a prerequisite for developing trustworthy AI, which in turn is frequently mentioned as both an ethical and regulatory goal independently of the specific field of deployment of the technology. As stated by Margrethe Verstager—Executive Vice-President for a Europe fit for the Digital Age in 2021 «[o]n artificial intelligence, trust is a must, not a nice to have.»[14] Moreover, particularly when AI is implemented in the healthcare sector to provide care to patients, such technology becomes part of a relationship founded on trust. This is due to the information asymmetry that exists between the patient and the doctor, with the latter being the only one able to fully comprehend a condition and potentially its treatment. Consequently, it is paramount for the patient to place her confidence in her physician concerning her health and well-being, as well as to adhere to the chosen care pathway that is most suitable for her specific clinical situation. Therefore, in order to preserve an adequate level of trust in the physician–patient relationship, every aspect of the provision of care should be covered, thus including the AI systems deployed to this end.

On the other, AI systems should be developed as *human-centered*, meaning that they should be used to enhance human capabilities, assist humans in achieving their goals, and, more generally, benefit mankind, instead of aiming at replacing them.[15] This leads to affirming ethical and legal principles that aim at maintaining humans in control of not only the development but also the deployment of the technology, especially in healthcare.

2.3 Issues related to AI ethics

As a matter of principle and of a general approach toward ethics, ethical principles should serve as an orientation for, or objectives of, one's actions. However, technological development generally considered, and not only when applied to healthcare, is mainly characterized by a tension between the interests of private companies investing in the research and development of AI, and the market on the one hand, both mainly driven by economic goals, and those of society at large on the other. The latter, while interested

---

[14] Juliette Faivre, «The AI act: towards global effects?», *European Science, Technology and Innovation Policy*, 2023, pp. 1–11.
[15] Mohamed Chetouani *et al.*, «Preface», in Mohamed Chetouani *et al.* (ed. by), *Human-centered artificial intelligence—advanced lectures*, Springer, 2023, pp. V-VIII.

in benefitting from the concrete application of the technology and the new possibilities it brings along, aims at the same time at protecting fundamental rights and values and thus fostering and pursuing a sustainable development of AI, which should thus be ethically sound.

A practical example of this tension is represented by a study conducted by McManara et al. in 2018, when the ethical debate on AI had been ongoing for decades, which showed that codes of ethics have almost no impact on software engineers and therefore on the practical side of the developmental process of AI systems.[16] Along the same line, in 2021 Gogol and colleagues argued that overall these principles are of «little immediate use during the software development processes,» leading to questionable practices such as cherry-picking («arbitrary accumulation of values» among the wide pool of those available in ethical guidance documents), indifference (not taking ethical principles under consideration because they lack concrete guidance and normative function), or ex-post orientation (ethical principles not identified to guide one's actions from the outset to reach an ethical goal, but chosen only after the desired goal has been reached and among those that adapt to such goal), to name a few.[17] Floridi has addressed these and other concerns among the «five risks of being unethical,» namely the already mentioned ethics shopping, ethics bluewashing, ethics dumping, ethics shirking, and ethics lobbying.[18]

Most of the mentioned issues arise from a non-clear, imprecise, or incoherent ethical framework and guidance,[19] and thus the ethical principles provided for in guidelines and documents run the risk of being ineffective[20] or having minimal effect on the desired purposes of contributing to guiding the development of a trustworthy and human-centered AI. Indeed, according to Munn, ethical principles on the matter are characterized as being *meaningless*, i.e., so highly abstract and ambiguous that they become incoherent, *isolated,* and thus not sufficiently connected to and

---

[16] Andrew McNamara, Justin Smith, and Emerson Murphy-Hill, «Does ACM's code of ethics change ethical decision making in software development?», In *Proceedings of the 2018 26th ACM Joint Meeting on European Software Engineering Conference and Symposium on the Foundations of Software Engineering*, Association for Computing Machinery, New York, NY, USA, 2018, pp. 729-733

[17] Jan Gogoll *et al.*, «Ethics in the Software Development Process: from Codes of Conduct to Ethical Deliberation», *Philosophy & Technology*, 34, 2021, pp. 1085-1108.

[18] Luciano Floridi, «Translating principles into practices of digital ethics: five risks of being unethical», *op. cit.* More extensively, Luciano Floridi, *Etica dell'intelligenza artificiale—Sviluppi, opportunità, sfide, op. cit.*

[19] Jessica Morley *et al., op. cit.*

[20] Thilo Hagendorf, «The ethics of AI ethics: an evaluation of guidelines», *Minds and Machines,* 30, 2020, pp. 99-120.

absorbed by the layers of society that actively and concretely contribute to the development of the field, and *toothless,* which refers to the absence of consequences that derive from the violation of said principles.[21] At the same time, different ethical principles may contrast with one another in a specific situation and the balancing exercise to be performed among them is frequently uncertain in both its methodology and result.[22] As O'Neill puts it—«There is no metric for balancing or trading-off different types of norms.»[23]

Indeed, the very nature of the ethical principles (at least) partly contributes to this result by being too abstract and value-laden, as well as not enforceable, and hardly concretely applicable in a field (that of AI) that is strongly process-oriented. This is mostly due to the fact that many of the guidelines developed on the matter adopted a principle-driven approach known as «principlism,» the aim of which is to provide high theoretical principles to guide both design practice and regulation.[24] This approach is traditionally the one adopted in bioethics and is nowadays at the core of some of the ethical approaches to AI.[25]

A possible solution to this issue is to include procedural or technical instructions among ethical principles, or conceptualizing them as such, in order to adapt to the concrete realm of developing AI systems,[26] thus attempting to translate the «*what* of AI ethics into the *how* of technical specifications.»[27] However, translational tools once again may present issues related to tool shopping (the choice of the one that best fits the concrete pathway already chosen by a company) and encourage «ethics by tick-boxing,» thus practical compliance with the tool without a real intention of ethically behaving.[28]

---

[21] Luke Munn, «The uselessness of AI ethics», *AI and Ethics,* 3, 2023, pp. 869-877.

[22] Anna Jobin, Marcello Ienca, and Effy Vayena, «The global land-scape of AI ethics guidelines», *Nature Machine Intelligence,* 1(9), 2019, pp. 389-99.

[23] Onora O'Neill, *From principles to practice: Normativity and judgement in ethics and politics,* Cambridge University Press, 2019.

[24] Alternative approaches have been identified as «realism» and «pragmatism». Rune Nyrup, Beba Cibralic, «Idealism, realism, pragmatism: three modes of theorising within secular AI ethics», in Barry Solaiman, and Glenn Cohen, *Research Handbook on Health, AI and the Law,* Edward Elgar Publishing, 2024, pp. 208–218.

[25] Luciano Floridi *et al.,* «AI4People—an ethical framework for a good AI society: opportunities, risks principles, and recommendations», *Minds and Machines,* 28, 2018, pp. 689-707.

[26] Thilo Hagendorf, *op. cit.*

[27] Jessica Morley *et al., op. cit.,* emphasis added. The concept is borrowed from Luciano Floridi, «Translating principles into practice of digital ethics: five risks of being unethical», *op. cit.*

[28] *Ibid.*

Nevertheless, the mentioned issues are more of «a danger to be avoided rather than as an inherent fault in the ethical turn».[29] Indeed, undoubtedly, ethical principles could not, and should not, lose their abstract character precisely because of their nature as *principles*, but they should nonetheless provide useful and concrete guidance for ethical conduct, especially in the process-oriented field of AI, thus finding the appropriate level of abstraction for their formulation and application.[30] At the same time, the relevant ethical principles so identified should be taken into consideration throughout the whole process of design, development, marketing, and use of AI, especially when applied to healthcare, according to an approach that could be defined as «Ethics by Design.» The latter concept is borrowed from that of Privacy by Design provided for by the European normative framework applicable to data protection and aims at underlying the need for developing AI in compliance with relevant ethical principles from the outset. To reach this goal, in 2021 Morley et al. proposed Ethics as a Service as a set of possible tools and measures to be adopted.[31] The overall idea behind this approach is that there are no «well designed or poorly designed algorithms» in terms of ethical acceptability or compliance with ethical principles, but the latter should be considered in any of the steps of the development of an AI system and by any of the people involved, therefore from the software engineers that technical design the technology to the personal who finally uses it to provide a service or more generally according to its intended purpose.[32] As a consequence, the approach to be adopted should be a collaborative one.[33]

The issue of the operationalization of AI Ethics and its principles has been tackled by various guidelines and documents on the matter, among which it is worth mentioning the Ethics Guidelines for Trustworthy AI published by the High-Level Expert Group on AI in 2019, the Recommendations of the OECD Council on Artificial Intelligence, and specifically for the healthcare sector the WHO Guidance—Ethics and Governance of Artificial Intelligence for Health of 2021. Most of these documents adopted a rights-based approach, which therefore tends to characterize the ethical principles established thereby as focused on the protection of the individual, leaving aside considerations about groups, communities, and society at

---

[29] Charles D. Raab, «Information privacy, impact assessment, and the place of ethics», *Computer Law & Security Review*, 37, 2020 pp. 1-16.

[30] Jessica Morley *et al.*, *op. cit.*

[31] *Ibid.*

[32] *Ibid.*

[33] Sabine Salloch, and Andreas Eriksen, «What are humans doing in the loop? Co-reasoning and practical judgment when using machine learning-driven decision aids», *The American Journal of Bioethics*, 24(9), 2024, pp. 67-78

large, which most of the time are barely mentioned.[34] However, AI has the potential to (possibly adversely) impact also society at large collectively and not only the patient who interacts with or is cared for by the technology, specifically when implemented within the healthcare sector. For instance, societal or collective concerns are related to exacerbated health inequalities, increased demands for services as access is rendered easier by technology, and environmental impact.[35] For this reason, other approaches have been proposed that include but also go beyond the individual by considering the impact of AI on groups and communities, institutions, the state, globally, and ultimately, over time.[36]

Over time, multiple studies have been conducted on the available guidelines and documents on AI ethics in order to establish the most cited principles as a common ground for further elaboration on the matter.[37] Moreover, the specificities of the healthcare sector, in terms of rights and values potentially at stake and the nature of the relationships involved, in particular the one between a patient and her physician, entail the need to tailor AI ethics to the field under consideration. As a consequence, the traditional ethical principles, as well as their adaptation to AI, should be considered through the lenses of the healthcare sector and its peculiarities. While the number of studies specifically focusing on AI ethics in healthcare is growing,[38] only a few documents or guidelines among all those devoted to the topic specifically address the issue, with special mention to the WHO Guidance, on the basis of which the following pages will be paragraphed.

Finally, on the matter, it should always be kept in mind that ethical principles are «dynamic» and «context-dependent,» and thus that they should be adapted and interpreted according to the circumstances under consideration.[39] In the specific field of healthcare, this means that the meaning of the ethical principles nowadays constituting AI Ethics, as well as the type

---

[34] Melanie Smallman, «Multi-scale ethics—why we need to consider the ethics of AI in healthcare at different scales», *Science and Engineering Ethics*, 28(63), 2022, pp. 1-17.

[35] Melanie Smallman, *op. cit.*

[36] *Ibid.*

[37] Anna Jobin, Marcello Ienca, and Effy Vayena, *op. cit.*; Jessica Fjeld *et al.*, *Principled artificial intelligence: Mapping consensus in ethical and rights-based approaches to principles for AI*, Berkman Klein Centre for Internet and Society, 2020; Thilo Hagendorf, *op. cit.*; Melanie Smallman, *op. cit.*

[38] Golnar Karimian, Elena Petelos, and Silvia Evers, «The ethical issues of the application of artificial intelligence in healthcare: a systematic scoping review», *AI and Ethics*, 2, 2022, pp. 539-551; Jessica Morley, and Luciano Luciano, «An ethically mindful approach to AI for healthcare», *Lancet*, 395, 2020, pp. 254-255.

[39] Onara O'Neill, *Constructions of reason: Explorations of Kant's practical philosophy*, Cambridge University Press, 1989.

and number of principles included therein, may change through time and according to developments of both the technology itself and the way healthcare is provided and (clinical, drugs, scientific, etc.) research is conducted. Moreover, these principles should also be interpreted «in light of each other,»[40] taking into account any change that possibly occurred.

2.4 Six relevant ethical issues

The WHO Guidance—Ethics and Governance of Artificial Intelligence for Health was published in 2021 and aims at providing «harmonised ethics guidance for the design and implementation of AI for global health,» addressing all levels of stakeholders involved throughout the healthcare system, from institutions to healthcare workers, patients, and software engineers.[41]

The document states at the very beginning of the ethical analysis that the proposed principles are grounded in the four «basic ethical requirements (...) that are considered noncontroversial»,[42] i.e., (1) *beneficence*, (2) *non-maleficence*, (3) *autonomy* and (4) *justice*, which can essentially be elaborated as (1) promoting (human) well-being, (2) doing no harm, (3) respecting for persons and their right, as well as ability, for deciding for themselves and thus making informed decisions about them, and (4) ensuring equitable and fair treatment of all people and avoiding any sort of discriminatory behavior. Quoting the work of Floridi and Cowls on the matter, when applied to AI development and use, the mentioned principles tackle the specificities of said technology, in particular that (1) AI should be designed and deployed for the benefit of humanity, and (2) potential misuse or overuse of the technology should be avoided.[43] At the same time, considering that we usually delegate to AI-specific tasks or the achievement of a given goal, it is also paramount (3) to strike «a balance between the decision-making power we retain for ourselves and that which we delegate to artificial agents» (i.e., «decide-to-delegate» model) and retain the right to modify such decision (i.e., deciding to decide again) and overall (4) AI should not increase discrimination and should actually be deployed for promoting justice.[44]

---

[40] Sabine Salloch, and Andreas Eriksen, *op. cit.*
[41] World Health Organization, *Ethics and governance of artificial intelligence for health.*, p. 19.
[42] World Health Organization, *Ethics and governance of artificial intelligence for health.*
[43] Luciano Floridi, and Josh Cowls, *op. cit.*
[44] *Ibid.*

To the four traditional principles, Floridi and Cowls suggested adding a fifth one, namely *explicability*, defined as both the possibility of truly understanding the functioning of the system, in relation to the issue of black-box (*intelligibility*) and that of allocating responsibilities for possible wrongdoings (*accountability*).[45]

This re-conceptualization of the four basic ethical principles to formulate a comprehensive theory of AI ethics has been later adopted as the foundation by the High-Level Expert Group on AI for the Ethics Guidelines for Trustworthy AI. Differently from the latter, whose aim was to identify and somehow institutionalized ethical principles generally applicable to AI, the WHO Guidance further declines them in the specific context of healthcare, as mentioned, and identifies the following principles: protect autonomy; promote human well-being, human safety, and the public interest; ensure transparency, explainability, and intelligibility; foster responsibility and accountability; ensure inclusiveness and equity; promote artificial intelligence that is responsive and sustainable.

2.4.1 Protecting autonomy

The ethical principle of the protection of autonomy is strictly related to most of the others included in the above-mentioned document, especially that of ensuring the transparency, explainability, and intelligibility of the system. While no clear, precise, and universally agreed-upon definition of *autonomy* can be provided,[46] in the specific context of AI ethics, it may be identified as the possibility for the human involved in the deployment of an AI system—be it a healthcare professional, a patient, or a researcher depending on the field of application—to retain a certain degree of control over the activities performed and the results delivered by the technology. This includes the ability of the clinician to guide or override the AI's medical decision or outcome, as well as that of the patient of self-determining in relation to one's health through the provision of informed consent and consent to data processing or other privacy-preserving measures patient.[47] The principle under consideration should also serve as a way to mitigate the so-called automation bias, i.e., the tendency to uncritically trust and adopt the output, the decision

---

[45] *Ibid.*

[46] Carina Prunkl, «Human autonomy in the age of artificial intelligence», *Nature Machine Intelligence*, 4, 2022 pp. 99-101.

[47] Generally on the issue, though not specifically for healthcare, Sábëlo Mhlambi, Simona Tiribelli, Decolonizing, «AI ethics: relational autonomy as a means to counter AI harms», *Topoi*, 42, 2023, pp. 867-880.

or the recommendation of an AI system,[48] usually caused by its high rate of success.

Furthermore, when AI is used for diagnosis or clinical care, AI should not only be designed to allow for potential human intervention at various stages—from input provision to output generation—but the system should also be transparent and comprehensible enough to ensure that the decision-making process and the algorithmic reasoning behind the output can be understood, and thus that the autonomy the physician is sufficiently preserved. The same line of reasoning should be adopted for the other areas of application of AI in the healthcare field, substituting clinicians with the appropriate person who should be in charge of such supervision. Concrete applications of this principle, or strategies, are expressed in the concepts of *human in the loop, human on the loop,* and *human in control*, thus more generally in ensuring an adequate level of *human oversight*.[49] However, it has been rightfully argued that the mentioned concepts are sometimes referred to as a «legal slogan» or a mere «source of ethical legitimacy without a clear delineation of which humans should be involved in which processes across the AI life cycle, and what those humans should be doing.»[50] Overall, these ethical principles foster a cooperative relationship between humans and AI systems, according to which, however, the decisive steps of the decision-making process are left to a human, also in the form of a mere supervision or possibility of overriding a decision already taken by the machine.[51]

Differently, when the perspective of the patient is adopted, *autonomy* does not perfectly overlap with *(informational) self-determination*,[52] but it can generally be translated into the need for the patient herself to be able to exercise an appropriate level of control over her own health by providing informed consent, or over their data processed by the AI system by providing consent to the processing. Various types of patient involvement and control are usually discussed, and in particular, it is currently under debate whether it is sufficient to safeguard their right to provide consent (in all its forms) or whether it would be more ethically sound to involve them as co-reasoners

---

[48] Markus Herrmann, «What are patients doing in the loop? Patients as fellow-workers in the everyday use of medical AI», *The American Journal of Bioethics*, 24(9), 2024, pp. 91-93.

[49] High-Level Expert Group on Artificial Intelligence, *Ethics Guidelines for Trustworthy AI*, 2019.

[50] Zachary Griffen, and Kellie Owens, «From "human in the loop" to a participatory system of governance for AI in healthcare», *The American Journal of Bioethics*, 24(9), 2024, pp. 81-83.

[51] Sabine Salloch, and Andreas Eriksen, *op. cit.*

[52] Sábëlo Mhlambi, and Simona Tiribelli, *op. cit.*

and fellow-workers in the design of the technology.[53] According to the authors, this would solve the problem of the principle of respect for autonomy being too vague.[54]

To successfully safeguard patients' autonomy in clinical decision-making processes involving AI, McDougall suggested designing AI systems to be value-flexible. He suggests creating machines that are not only sensitive to shared values in societies but also to the specific set of personal values of the patient under consideration. Indeed, not every clinical decision or pathway is appropriate for every patient, even in similar clinical situations, precisely because of their right (and duty) to choose what is best for them. As a consequence, and to preserve true patient-centered healthcare, both sets of values (societal and personal of the single patient) should be considered and included in the design of the technology to be used. The assessment of the appropriate and the best way to collect such individual preferences, and therefore if using databases, historical data, social media, or other means, is, however, a challenging task.

Finally, attempting at defining what autonomy should look like in the medical decision-making process from the point of view of the patient, the question of whether the latter has a right to refuse treatments based on, or that otherwise involve, AI should be addressed. This right may be conceptualized as both a positive right, i.e., the right of the patient to choose not to be subject to a diagnosis performed, or a treatment planned by AI, and a negative right, and thus pertaining to her possibility of insisting on receiving an alternative type of healthcare assistance.[55] Once again, the matter is still under debate. While usually answers are positive regarding the abstract legitimacy of its introduction, questions remain about its nature.[56] Indeed, two versions of the right have been identified: a strong version, according to which a patient would have the right to refuse any involvement of AI in her medical journey whatsoever, and a weak version, that would give the patient to possibility of asking for a physician to be involved alongside with AI.[57] It remains to be seen whether a right to refuse to be subject to automated medical decision-making will be introduced by law.

---

[53] Sabine Salloch, and Andreas Eriksen, *op. cit.*

[54] *Ibid.*

[55] Holm Plou, «The right to refuse diagnostic and treatment planning by artificial intelligence, medicine», *healthcare and Philosophy*, 23, 2020, pp. 107-114.

[56] On the affirmative side of its recognition, Holm Plou, *op. cit*; Iñigo de Miguel Beriain, «Should we have a right to refuse diagnostics and treatment planning by artificial intelligence?», *Medicine, healthcare, and Philosophy*, 23(2), 2020, pp. 247-252.

[57] Iñigo de Miguel Beriain, *op. cit.*

2.4.2 Promoting human well-being, human safety, and the public interest

As previously mentioned, ethical AI should be developed as human-centered and therefore to promote human well-being, by ensuring the safety, accuracy, and efficacy of the technology throughout the whole life cycle of the system. In this sense, it is strongly related to the «traditional» ethical principle of non-maleficence.

This principle is likely the most overarching, aiming to ensure that AI is developed for the benefit of humanity and its welfare, rather than to replace humans or cause harm. Given the dual potential of AI systems to provide significant benefits while also posing risks, all related activities must be guided by this fundamental principle. At the same time, however, it is also the most challenging to accurately define and ensure its applicability, as it is subject to change over time in response to developments in both society and technology.

2.4.3 Ensuring transparency, explainability, and intelligibility

While the WHO Guidance lists *transparency*, *explainability*, and *intelligibility* separately, individually defining the terms and identifying the differences among them may not be straightforward.[58]

Indeed, the principle of explainability can be understood as an umbrella term[59] that includes all of those mentioned above, and its main aim is to ensure that AI is intelligible and/or understandable both in terms of its reasoning and its output to whoever is involved in its design, development, and use, from scientific engineers to regulators.[60] The need for complying with this ethical principle to ensure safe and robust healthcare is frequently addressed in the literature as one of the most pressing needs (if not the preeminent one).[61]

---

[58] Brent Mittelstadt, Chris Russell, and Sandra Wachter, «Explaining Explanations in AI», in *Proceedings of the Conference on Fairness, Accountability, and Transparency*, 2019, pp. 279-288.

[59] Borrowing the conceptualization developed for explicability by Scott Robbins, «A misdirected principle with a catch: Explicability for AI», *Minds and Machines*, 29, 2019, pp. 495-514; Frank Ursin, Cristian Timmermann, and Florian Steger, «Explicability of artificial intelligence in radiology: is a fifth bioethical principle conceptually necessary?», *Bioethics*, 36(2), 2022, pp. 143-153.

[60] Kawamleh talks about «explainability imperative» when AI is deployed in «safety critical domains like healthcare». Suzanne Kawamleh, «Against explainability requirements for ethical artificial intelligence in healthcare», *AI and Ethics*, 3, 2023, pp. 901-916.

[61] Among many others, Zachary C. Lipton, «The mythos of model interpretability: In Machine learning, the concept of interpretability is both important and slip-

It is precisely due to the significance of developing a trustworthy AI for the society of the mentioned characteristics (transparency, explainability, and intelligibility) that a variant of them has been proposed to be included among the five traditional principles of bioethics, namely explicability.[62] The latter has been defined as «the crucial missing piece of the AI ethics jigsaw» and the sum of both intelligibility and accountability.[63] Contrary to this approach, Ursin et al. claim that there is no need for adding a fifth principle to those originally identified by Beauchamp and Childress and that explicability should be rather considered as a means for ensuring compliance with the non-maleficence principle.[64] Providing a solution to this debate is beyond the scope of this work. However, it underlines the importance for society to understand how an AI system has reached its outcome in order to fully trust it.

The WHO Guidelines provide for two different approaches to this end: (1) improving the transparency of the technology by providing the public and stakeholders sufficient information on the functioning of the AI system, (2) guaranteeing its explainability, i.e., the possibility of understanding the steps in the pathway chosen by the system to reach a certain output.

On the one hand, ensuring transparency is essential for healthcare professionals to improve the quality of healthcare and protect patient's health by precisely delimiting the purpose of the technology, its strengths, and limitations. On the other, explainability should be tailored to the capacity and knowledge of the intended user (healthcare professionals) or recipient (patient) population. This second conceptualization of the principle resembles the right to explanation and the corresponding duty to provide accurate, precise, and tailored information already provided for by the data protection regulation, with which it also shares the ratio.[65]

As such, explainability is strictly related to, and sometimes it is a prerequisite for, ensuring respect for the other mentioned ethical principles.[66]

---

pery», *Queue* 16(3), 2018, pp. 31-57; Danton S. Char, Michael D. Abràmoff, and Chris Feudtner, «Identifying ethical considerations for machine learning healthcare applications», *The American Journal of Bioethics*, 20(11), 2020, pp. 7-17. On the matter, Robbins argues against «a principle requiring that AI be explicable», even though he admits that explicable AI could be useful in certain contexts. However, among these contexts, the author includes healthcare, for the provision of which it would be «wrong» to deploy inexplicable AI. Scott Robbins, *op. cit.*

[62] Luciano Florici, and Josh Cowls, *op. cit.*

[63] *Ibid.*

[64] Frank Ursin, Cristian Timmermann, and Florian Steger, *op. cit.*

[65] Scott Robbins, *op. cit.*

[66] Alexander Kempton, and Polyxeni Vassilakopoulou, *Accountability, Transparency & Explainability in AI for Healthcare*, 8th International Conference on Infrastructures

Indeed, for instance, the autonomy and self-determination of a healthcare professional or patient can be ensured only insofar as it is possible for them to truly understand the functioning and the reasoning pathway of the AI system. Moreover, explainability enables meaningful human control over the technology itself, i.e., «giving humans the ability to accept, disregard, challenge, or overrule an AI algorithm's decision».[67] Finally, it could help allocate responsibility among the people and the AI systems involved, because it would make clear who caused or generated a given outcome and thus should be held accountable for it.

Against compliance with this principle runs the issue of black-box AI, i.e., when the inner workings of such technology cannot be grasped or fully understood (and thus when it is *opaque*), not even by experts in the field. This is particularly troublesome in healthcare where decisions with an impact on the clinical pathway of a patient may be taken by the healthcare professionals partly relying on a technology that they cannot completely understand. «Relying on devices whose logic is opaque violates principles of medical ethics»[68] and it «seems to be in direct conflict with the right of patients to be provided with meaningful information about the logic involved as well as about the significance and the envisaged consequences of diagnostic or treatment procedures, or other interventions into health (Bachle 2019)».[69]

Finally, the principles discussed in this paragraph are closely related to two other issues. Firstly, *explainability* tends to exist at opposite ends of the spectrum from *accuracy* in complex AI systems, meaning that enhancing explainability may come at the cost of accuracy. Consequently, one often faces a trade-off between achieving complete explainability in an algorithm, on the one hand, or enhancing its accuracy (and therefore entailing a certain level of opacity) on the other.[70] Such a trade-off can only be performed according to the specific circumstances of a given application of AI, because «a blanket preference for simpler models [to be read as *more explainable*] is simply a lethal prejudice».[71]

---

in Healthcare, 2021; Julia Amann *et al.*, «Explainability for artificial intelligence in healthcare: a multidisciplinary perspective», *BMC Medical Informatics and Decision Making*, 20(310), 2020, pp. 1-9.

[67] Scott Robbins, *op. cit.*

[68] Shinjini Kundu, «AI in medicine must be explainable», *Nature Medicine*, 27(8), 2021, pp. 1328-1328.

[69] Suzanne Kawamleh, *op. cit.*

[70] World Health Organization, *Ethics and governance of artificial intelligence for health*; Frank Ursin, Cristian Timmermann, and Florian Steger, *op. cit.*

[71] Alex J. London, «Artificial intelligence and black-box medical decisions: accuracy versus explainability», *The Hastings Center Report*, 49(1), 2021, pp. 15-21.

Furthermore, since explainability involves possessing and sharing insights into an AI system's operations, it is important to identify *who* qualifies to receive this information, particularly whether the patient should be included. Consequently, the ethical principle of explainability must be evaluated alongside informed consent. Regarding this issue, Scott argues that details about how the system operates should only be disclosed to the individual who operates the AI system, not to those who are indirectly impacted by its outcome.[72]

2.4.4 Fostering responsibility and accountability

As technology develops, AI systems in healthcare are performing progressively more (and more complex) activities. Thus, they frequently substitute healthcare professionals in performing specific tasks that were the latter's exclusive responsibility until then. This, in turn, raises the issue of correctly allocating responsibilities and accountabilities in the event of a mistake, malfunction, or wrongdoing that adversely impacts the patient. The significance of civil liability rules was highlighted in the World Health Organization's 2024 report «Ethics and governance of artificial intelligence for health: guidance on large multi-modal models».[73]

First and foremost, for the purposes of this discussion, we should clarify that accountability and liability are not interchangeable terms. Accountability is more closely associated with a societal duty, particularly toward patients, whereas liability often pertains to legal obligations, which can vary depending on the specific legal frameworks in place. However, considering that human healthcare providers might have limited influence over choices made by AI systems and may lack deep insight into the decision-making processes of these technologies, addressing both accountability and the complexity of liability becomes problematic when harm to patients occurs.

The implementation of AI systems for the provision of care may give rise to a «diffusion of responsibility,» i.e., a situation in which the involvement of different actors (such as, in the case under consideration, the healthcare professional, the AI system itself, the provider of the technology, the researcher, etc.) in instances of damage, to patients or otherwise, may result in the allocation of responsibility for such damage being «manifold, uncertain, or not

---

[72] Scott Robbins, *op. cit.*

[73] World Health Organization, *Ethics and governance of artificial intelligence for health: guidance on large multi-modal models*, https://iris.who.int/handle/10665/375579, 2024.

consolidated in particular administrative, legal, or social structures».[74] In this situation, it may well be the case that no one would be held accountable at all, especially as technology evolves and becomes even more autonomous in the tasks performed and the decisions taken.[75]

Consequently, it is necessary to adequately provide «clear and transparent specifications»[76] on what a given AI system may concretely achieve and under what circumstances. This way, the healthcare professional or more generally the user could delimit the scope of application of AI in the clinical decisions to be taken, provide appropriate input and require appropriate output, as well as be aware of the relationship between their autonomy and that of the AI. The mentioned specifications should be provided by the manufacturer of the system, who is not only the stakeholder closer to the information itself but also the one that identifies the perimeter of the functioning of the AI in the first place. In turn, in the words of the WHO Guidance, allocating tasks means allocating responsibilities, in the sense that by precisely delimiting tasks of all the stakeholders involved it is also possible to trace back the action that has a causal link to the damage caused. As it appears evident, compliance with this principle may only be ensured if the system is adequately explainable or intelligible both to the provider and the user, and as a consequence, the two ethical principles are closely related to one another.

At the same time, adequate accountability measures should be in place for redressing individuals and groups adversely affected by a medical decision taken by the healthcare professional with the aid of an AI system. Such redress should include «compensation, rehabilitation, restitution, sanctions where necessary, and a guarantee of non-repetition».[77] To this end, the WHO proposed to adopt the model of «collective responsibility», according to which «all agents involved in the development and deployment of an AI technology are held responsible» with the final aim of «encouraging all actors to act with integrity and minimise harm».[78]

---

[74] Bleher Braun, and Hannah Matthias, «Diffused responsibility: attributions of responsibility in the use of AI-driven clinical decision support systems», *AI and Ethics*, 2, 2021, pp. 747-761.

[75] Of the opposite view, Daniel W. Tigard, «There is no techno-responsibility gap», *Philosophy & Technology*, 34, 2020, pp. 589-607.

[76] World Health Organization, *Ethics and governance of artificial intelligence for health*.

[77] *Ibid.*

[78] *Ibid.*

2.4.5 Ensuring inclusiveness and equity

In the healthcare sector, it is paramount to ensure that decisions with direct or indirect consequences on patients and the provision of healthcare are sufficiently inclusive and fair. Failure to consider or address parts of the population may result in biased or discriminatory outputs and medical decisions, as well as unequal treatments and the consequences on patients' rights may be enormous and unpredictable. Among the applications of AI in healthcare with direct consequences on patients, reported examples of unfairness or biased results include an algorithm used to identify patients who deserved special medical attention that significantly discriminated against racial minorities because of existing access barriers to healthcare for them,[79] and AI tools to detect Alzheimer's disease by analyzing a sample of the speech of a patient, and that however was able to correctly perform only for people of a specific Canadian dialect.[80] [81]

Indeed, AI systems in healthcare are «especially susceptible to bias and discrimination»[82] and it is nowadays notorious that they may provide biased outcomes for various reasons, which have been grouped by Rajkomar and colleagues into bias in model design, in data (especially training data), in interactions with clinicians and with patients.[83] Collectively, they may be referred to as *intrinsic unfairness* which affects the results of AI[84] and may happen in any of the actual possible applications of AI in healthcare. These biases frequently derive from social inequalities and actual discriminatory practices occurring in the ordinary provision of healthcare (so-called feedback loop bias[85]), from which most of the time training data are collected and in which AI systems are introduced, as well as because of the existence

---

[79] Ziad Obermeyer *et al.*, «Dissecting racial bias in an algorithm used to manage the health of populations», *Science*, 366(6464), 20019, pp. 447-453.

[80] Dave Gershgorn, «If AI is going to be the world's doctor, it needs better textbooks», 2018, available at https://qz.com/1367177/if-ai-is-going-to-be-the-worlds-doctor-it-needsbetter-textbooks

[81] For a more in depth description of examples of AI biased results in healthcare, Sharona Hoffman, and Andy Podgurski, «Artificial intelligence and discrimination in healthcare», *Yale Journal of Health Policy, Law, and Ethics*, 19(3), 2020, pp. 1-49.

[82] Daniel Schönberger, «Artificial intelligence in healthcare: a critical analysis of the legal and ethical implications», *International Journal of Law and Information Technology*, 27(2), 2019, pp. 171-203.

[83] Alvin Rajkomar *et al.*, «Ensuring fairness in machine learning to advance health equity», *Annals of Internal Medicine*, 169(12), 2018, pp. 866-872.

[84] Matthew Fenech, Nika Strukelj, and Olly Buston, *Ethical, Social, and Political Challenges of Artificial Intelligence in Health*, Future Advocacy, Wellcome Trust, 2018.

[85] Sharona Hoffman, and Andy Podgurski, «Artificial Intelligence and Discrimination in healthcare», *op. cit.*

of underrepresented categories of patients (so-called selection bias[86]). The latter often include racial minorities, older patients, patients with multiple morbidities, pregnant women, and children. For any of the mentioned groups, data to be used for training AI systems may be lacking for historical reasons and existing social inequalities, difficulties in acquiring informed consent for the medical procedure, ethical reasons, limited or insufficient number of patients overall or in a given area, underrepresentation, etc.[87] Indeed, biased AI outcomes may derive from intrinsic biases in the dataset used to train the AI system and in the input data provided by the user of such technology. For instance, in 2019 a commercial prediction algorithm trained using data about the healthcare costs of various patients, from which race was excluded in order to avoid biases, resulted in seriously biased outcomes precisely because care was not a race-neutral measure.[88]

At the same time, even if *intrinsic unfairness* is absent, discriminatory outcomes may derive from the use of unbiased AI systems, i.e., what we can address as *extrinsic unfairness*, which would limit access to fair healthcare.[89] This may occur because of various factors. First of all, there may be an unfair use of the technology by healthcare professionals (consciously or unconsciously), who may treat with an AI system patients that are not the correct target of the chosen technology. Moreover, through time the situation may further be exacerbated because of the effects of the so-called digital divide, defined by the OECD as «the gap between individuals, households, businesses and geographic areas at different socioeconomic levels with regard both to their opportunities to access ICT and to their use of the Internet for a wide variety of activities.»[90] Indeed, as a matter of example, it may be challenging for older patients to use, and completely trust, mobile applications or smart devices to manage their health or for correctly adhering to their clinical pathway.

To avoid biases of the first and second kind, it is suggested to include not only in the design and development but also actual use of the AI systems people from various races, ages, backgrounds, and cultures, as well as patient

---

[86] Sharona Hoffman, and Andy Podgurski, «The use and misuse of biomedical data: is bigger really better?», *American Journal of Law & Medicine*, 497, 2013, pp. 521-523.

[87] Siân Care, Allan Pang, and Marc de Kamps, «Fairness in AI for healthcare», *Future Healthcare Journal*, 11(3), 2024, pp. 1-3.

[88] Stuart McLennana *et al.*, «AI ethics is not a panacea», *American Journal of Bioethics*, 20(11), 2020, pp. 20-22.

[89] Matthew Fenech, Nika Strukelj, and Olly Buston, *op. cit.*

[90] OECD, «Understanding the digital divide», *OECD Digital Economy Papers*, No. 49, OECD Publishing, Paris.

representatives, and take their data, experiences and opinions under consideration throughout the whole life-cycle of the AI system. Moreover, the specific set of data to train the algorithm should be carefully evaluated by the provider in terms of inclusiveness and double-checked, as much as feasible, for its appropriateness by healthcare professionals (or healthcare facilities) when choosing the technology to use for a specific patient. As mentioned, the principle of inclusiveness and equity also necessitates addressing the digital divide to avoid unequal access to AI systems and AI-driven healthcare across various states, and healthcare facilities as well. Indeed, ultimately, a conscious use of AI would not only provide unbiased and not-discriminatory outputs, but it may also be used to actively reduce existing disparities between stakeholders and patients, as well as human biases usually performed in the provision of care.[91] After all, the aim is not to create and use a technology «calibrated for younger, more urban bodies».[92]

2.4.6 Promoting artificial intelligence that is responsive and sustainable

Especially in the field of healthcare, it is paramount to avoid misdiagnosis and errors, which may be caused by either flaws in the technology or the physician's over-reliance on the output/response of the system (and sometimes both). As a consequence, the principle of responsiveness is crucial, linked to the need for «developers, policymakers, and healthcare practitioners to systematically and continuously evaluate AI systems»[93] to ensure that the technology is fit for purpose, taking into consideration the state of the art and also the specific context in which it is destined to operate.

As for the principle of sustainability, AI systems should be developed by carefully considering their possible environmental impact. While the concrete impact of this technology on the environment is yet to be evaluated, some worrisome data are already known. For instance, the training of Google's AlphaGo Zero game for 40 days had the same carbon impact of 1,000 hours of air travel,[94] and more generally the extraction of minerals and

---

[91] Milena A. Gianfrancesco *et al.*, «Potential biases in machine learning algorithms using electronic health record data», *178 Jama International Medicine*, 2018, pp. 1544-1546.

[92] Robert David Hart, «If you're not a white male, artificial intelligence's use in healthcare could be dangerous», https://qz.com/1023448/if-youre-not-a-white-male-artificial-intelligences-use-in-healthcare-could-be-dangerous.

[93] World Health Organization, *Ethics and governance of artificial intelligence for health*.

[94] For a more in depth study on AI's carbon emission see Kate Crawford, *The atlas of AI: Powers, politics, and the planetary costs of artificial intelligence*, Yale University Press, 2021.

metals, as well as the creation of plastics necessary for building AI components, is known to have a tremendous impact on the environment.[95] More
generally, AI should be «not just technically and clinically viable, but also
socially, ethically, and environmentally accountable».[96] Indeed, it has been
shown that AI is energy-intensive both in its training and deployment and
generates high quantity of electronic waste, and that data centers contribute
to greenhouse gas emissions.[97]

Consequently, the principles of environmental and systemic sustainability should nonetheless be carefully taken into consideration along the way
so that irreversible consequences can be avoided or at least mitigated and
preventive and corrective measures can be adopted. In this sense, AI should
be both developed and designed (sustainable development), as well as used
(sustainable use) in order to limit its negative consequences on the environment, even though the dividing line between the two stages in the life-cycle
of an AI system is not always clear.[98] Applying the principle of sustainability
of AI would mean choosing ethically sound projects to develop and applications, in spite of others equally useful for society but more dangerous for
the environment.

At the same time, however, it should also be considered that AI systems
may be implemented in healthcare to reduce the environmental impact of
healthcare facilities or the healthcare system as a whole, for instance by optimizing workflow, reducing waste and facilitating telemedicine or other low
environmental impact practices.[99]

2.5 Legal issues

Alongside the ethical issues, AI raises various legal concerns as well. As
mentioned, these are partly influenced by the former but differ from them
primarily due to their enforceable nature. Following the categorization provided for by Gerke et al.,[100] these issues may be grouped as: (I) safety and

---

[95] Cristina Richie, «Environmentally sustainable development and use of artificial
intelligence in healthcare», *Bioethics*, 36, 2022, pp. 547-555.

[96] World Health Organization, *Ethics and governance of artificial intelligence for
health*.

[97] Daiju Ueda *et al.*, «Climate change and artificial intelligence in healthcare: review and recommendations towards a sustainable future», *Diagnostic and Interventional Imaging*, 105(11), 2024, pp. 453-459.

[98] Cristina Richie, *op. cit.*

[99] Daiju Ueda *et al.*, *op. cit.*

[100] Sara Gerke, Timo Minssen, and Glenn Cohen, «Chapter 12—Ethical and legal challenges of artificial intelligence-driven healthcare», in Adam Bohr, and Kaveh

effectiveness; (II) liability; (III) privacy and data protection; (IV) cyberse-curity; (V) intellectual property law. They will be analyzed in the following pages.

Any of these issues are conceptually linked to fundamental rights or val-ues potentially endangered by the deployment of AI, in general and within the healthcare sector, as protected under national and international norms. To address them, any legislator has to face the choice of either adjusting existing norms by way of interpretation to cover the newly raised issues or enacting new provisions specifically tailored for the new instances of pro-tection. Safeguarding rights and values is indeed one of the justifications for regulating new technologies. However, this should be balanced against the counterarguments in favor of not regulating AI (or any other new technol-ogy) in order to avoid «constrain[ing] innovation, impos[ing] unnecessary burdens, or otherwise distort[ing] the market».[101] The difference between these two approaches, i.e., *ex ante* regulating new technologies or any activi-ty potentially endangering human rights and fundamental values as opposed to refraining from doing so in order to allow the maximum level of market freedom, is mirrored in the approaches adopted by the European Union, usually defined as *human-right based*, and that of the United States, com-monly addressed as *market-driven*.

Moreover, any regulation that aims to address new technologies faces the so-called Collingridge dilemma. According to this theory first developed in 1980, norms to be enacted in this field have their maximum potential for effectively controlling a new technology at the beginning of its development, when however little is known about the potentially harmful consequences of its development and use.[102] The solution proposed in this regard is not to focus on preventing harm but on ensuring that norms are future-proof and technology-neutral. In this regard, two approaches have been identified as relevant: the precautionary principle and masterly inactivity.[103]

As for the first one, adopting the precautionary approach aims to guide decisions whenever uncertainties regarding the causes of possible risks gov-ern the matter. In this case, Member States may «take protective measures without having to wait until the existence and gravity of those risks become

---

Memarzadeh (ed. by), *Artificial intelligence in healthcare*, Academic Press, 2020, pp. 295-336.

[101] Simon Chesterman, «Chapter 8: From ethics to law: why, when, and how to regulate AI», in David J. Gunkel (ed. by), *Handbook on the Ethics of Artificial Intelli-gence*, Edward Elgar Publishing, 2024, pp. 1-22.

[102] David Collingridge, *The Social Control of Technology*, Frances Pinter, 2020. On this, see also Simon Chesterman, *op. cit.*

[103] *Ibid.*

fully apparent,» but at the same time, they must base the risk assessment on concrete facts and not purely hypothetical considerations.[104] Concerning the second strategy, it entails a passive approach of delaying action on the issue by withholding the implementation of regulations until deemed necessary.

Furthermore, it is argued that a regulation on AI should be «universal and not domain specific» because of the «far reaching consequences of AI».[105] While the domain-specificities should not be ignored by any attempt at regulating the technology, a universal and horizontal approach should be preferred, in order to guarantee a higher degree of harmonization and protection.

Ultimately, it is important to remember that when dealing with the legal matters outlined in the subsequent sections, lawmakers often need to strike a balance between conflicting interests and the very issues they aim to address. For instance, ensuring an adequate level of transparency of the technology entails possible trade-offs with cybersecurity, accuracy, privacy, and intellectual property.[106]

2.5.1 Five relevant legal issues—(I) safety and effectiveness

First of all, especially in healthcare, it is paramount to ensure that AI systems used to provide care for patients, to assist physicians, or otherwise in the field are *safe and effective*. This means that standards are needed to guide the design and development of such technologies in order to reasonably ensure the achievement of the desired (clinical) result while simultaneously not causing harm or damage to patients, healthcare providers, and users.

This goal can be reached by «making sure that the datasets are reliable and valid, performing software updates at regular intervals, and being transparent about [the] product, including shortcomings such as data biases,» as well as testing the technology in order to ensure that it can reach the aim for which it is designed (or intended use) with an adequate level of effectiveness and the minimum level of risks involved. Consequently, norms should be enacted to ensure that the «highest safety standards available»[107] are complied with in the design, production, and deployment of

---

[104] CJEU, Case C-41/02, Commission v. Netherlands, para 52.

[105] Nicolas P. Terry, «Of regulating healthcare AI and robots», *Yale Journal of Law & Technology (YJoLT)*, 21, 2019, pp. 1-53.

[106] Cohen Glen *et al.*, «The European artificial intelligence strategy: implications and challenges for digital health», *The Lancet Digital Health*, 2(7), 2020, pp. 376-379.

[107] Eduard Fosch Villaronga, *op. cit.*

AI. Implementing the ethical principle of autonomy, these norms should also establish an «adequate level of oversight».[108]

At the same time, legal rules should also address the issue of updates or modifications of AI systems after they have been placed on the market, so-called *update problem*.[109] Indeed, many AI systems are designed to update or be updated through time and use. The provider may directly perform some of these modifications, while others may not have been envisioned at the time of making the system available for use. Locking the algorithm[110] at the moment of its first commercialization, i.e., preventing such autonomous modifications from happening can provide a potential solution to this issue. However, at the same time, it eliminates many of the advantages associated to the use of AI systems (especially deep learning) and not other types of software or technologies. As a consequence, instead of prohibiting updates of this kind, it is advisable that norms on the matter are designed to «continuously monitor, identify, and manage risks associated with these algorithms» through time.[111] The aim in this regard would be to ensure that potential updates or modifications of the AI system do not impair the overall safety of the technology or its ability to be used as intended.

Finally, safety should not be seen as an absolute requirement inherent to the AI system and independent of the context where it is implemented; instead, it should depend on it and on the intended use of the technology. As a consequence, AI systems that function safely and effectively when used as intended by the provider may become less safe if used for purposes outside those specified, and for which the AI has been designed. Therefore, aligning with the principle of transparency, norms should prescribe that information about the AI system is provided in a form that it is possible for deployers not only to clearly understand what the system does, under which conditions and for what purposes but also how the system is not designed to perform. Its capabilities, as well as its limitations, are fundamental information that should be provided to users (healthcare professionals, patients, researchers etc.) to ensure the safety and effectiveness of the use of the AI system.

---

[108] Sara Gerke, Timo Minssen, and Glenn Cohen, *op. cit.*

[109] Cohen Glen *et al.*, *op. cit.*

[110] Boris Babic *et al.*, «Algorithms on regulatory lockdown in medicine», *Science*, 366, 2019, pp. 1202-1204.

[111] Cohen Glen *et al.*, *op. cit.*

2.5.2 Five relevant legal issues—(II) liability

Moreover, due to its intrinsic characteristics, AI poses significant challenges to the liability frameworks already in force around the world. In particular, when complex systems are deployed for healthcare purposes, it can become especially troublesome, or in certain instances impossible, to establish the causal link between an action, the ensuing damages, and the accountable entity, whether human or technological. This becomes particularly noticeable if black-box AI is implemented and in situations where the system can learn from experience and update through time. Indeed, in the first scenario, if it is not feasible to comprehend the reasons behind the system's output, it becomes equally problematic to allocate responsibilities. The same reasoning applies if the AI system is opaque enough to prevent physicians from predicting possible erroneous outcomes.[112] Moreover, as for the second, the (legal) dilemma to be solved relates to the justification for considering the human (provider or healthcare professional) responsible for damage caused by the autonomous behavior of the machine,[113] which was not specifically envisioned at the time of making it firstly available for use.

In particular, current legal regimes might not be adequate for regulating the use of AI in healthcare, where multiple entities are involved, some of which may be completely new to the field. Indeed, in these instances, the provider of the system, the healthcare professional or the researcher and the technology itself all contribute to delivering a certain outcome (and therefore, in the case under consideration, a damage) to various extents and with different levels of autonomy and causality. Multiple scenarios may be envisioned in this regard, two of which are particularly challenging for liability regimes.[114] On the one hand, the AI system might correctly advise a care pathway that deviates from the standard of care, and the physician might wrongfully decide not to follow the AI outcome, causing damage to the

---

[112] Eduard Fosch Villaronga, *op. cit.*

[113] In support of the idea not to consider the humans responsible in this case, Andreas Matthias, «The responsibility gap: ascribing responsibility for the actions of learning automata», *Ethics and Information Technology*, 6(3), 22004, pp. 175-183; Deborah G. Johnson, «Technology with no human responsibility?», *Journal of Businell Ethics*, 127(4), 2015, pp. 707-715. Of the opposite view, Jatinder Singh, Christopher Millard, and Jennifer Cobbe, «Accountability in the IoT: systems, law and ways forward», *Computer*, 51(7), 2018, pp. 54–65; Jatinder Singh *et al.*, «Responsibility & machine learning: part of a process», *Social Science Research Network*, 2020, pp. 1-20.

[114] For an overview of these scenarios, see Price II Nicholson, Gerke Sara, and Cohen Glenn, «Liability for use of artificial intelligence in medicine,» in Barry Solaiman, Glenn Cohen (ed. by), *Research Handbook on Health, AI and the Law*, Edward Elgar Publishing, 2024, pp. 150-166.

patient. The other case refers to the decision of the physician to follow such advice from the AI system, which, however, is incorrect in this instance. In both cases, the physician with her decision somewhat causes damages to the patient, but whether or not it is advisable to hold her responsible for such action remains debatable. On top of this, regulations should also address potential responsibilities of hospitals or healthcare facilities directly toward patients.

Determining the extent of liability for physicians and healthcare facilities also impacts the level of development and implementation of AI technologies in current medical practice,[115] as it may well be the case that a new technology is not adopted if risks of being held responsible for its autonomous outcomes are too high or undeterminable. The issue should thus be carefully addressed by legislators in order not to indirectly restrain technological development from occurring, and ensure an adequate level of clarity.

2.5.3 Five relevant legal issues—(III) privacy and data protection

The amount of data processed in the context of AI, whether for training the system or for subsequently providing the desired and intended outcome, calls for a rigorous system in place to adequately protect the rights to privacy and data protection of the data subjects involved. In the specific context of healthcare, this would entail regulating the processing of «ordinary» as well as sensitive data, such as health and genetic data, and adjusting any norm or principle to the realm of data processing in the context of AI, which usually involves large quantity of data, sometimes at the level of big data. Indeed, as addressed in the previous pages, AI systems need high quantity of data to ensure that they are designed free from unwanted biases and discriminatory patterns and nowadays the healthcare field as a whole is progressively being populated by mobile medical apps, wearables devices, chatbots, and connected devices, all of which constantly collect and processed personal data. Moreover, precisely to avoid biases in the data sets used to train the algorithms, some of these data are collected from, or pertain to vulnerable categories. At the same time, these data should be accurate and fit-for-purpose[116] to avoid discriminatory outcomes, especially caused by statistical biases of the algorithm.[117]

---

[115] George Maliha *et al.*, «Artificial intelligence and liability in medicine: balancing safety and innovation», *The Milbank Quarterly*, 99(3), 2021, pp. 629-647.
[116] Eduard Fosch Villaronga, *op. cit.*
[117] Ravi B Paikh, Stephanie Teeple, and Amol Navath, «Addressing bias in artificial intelligence in healthcare», *Jama*, 322(24), 2019, pp. 2377-2378.

Risks in this regard are related, among others, to the «leak of sensitive information through data breaches, personal data transfer, and sale to third parties.»[118] Moreover, a fundamental contrast should be addressed by norms on the matter, and in particular the quasi-ontological differences between the basic principle of most of the regulation on data processing, i.e., data minimization, whose aim is to safeguard the right to data protection and privacy, and AI which flourishes thanks to large amounts of data.[119]

Finally, it should also be regulated the possibility of reusing for different purposes data previously collected for training and testing a specific AI system, for instance with the aim of updating it or for statistical purposes. Indeed, secondary uses of this kind may serve the goal of ensuring the appropriate level of safety of the technology, but if not appropriately regulated may run the risk of infringing patients' fundamental rights.

2.5.4 Five relevant legal issues—(IV) cybersecurity and (V) intellectual property law

Intrinsically linked to the issue of safety and data protection is that of cybersecurity. Indeed, in an (inter)connected world, of which AI systems constitute only one of the multiple components,[120] risks derived from breaches of the security of these systems may have unprecedented consequences. Indeed, especially with the advent of the Internet of Things and wearable devices, the dividing line between the physician and the online world has become progressively blurred, thus increasing also the scale of potential harms if an adequate level of cybersecurity of these technologies is not guaranteed. Threats may derive from both external attacks on the connected devices used in the field and internal errors of the people in charge of using them,[121] and have drastically increased through time[122] with the progressive digitization of healthcare, for instance with the advent of the Electronic Health Records (EHR) that brought patients' sensitive data on the online world.

Indeed given the importance of the values to be protected, scholars start-

---

[118] Eduard Fosch Villaronga, *op. cit.*

[119] Tom Sorell, Nasir Rajpoot, and Clare Verrill, «Ethical issues in computational pathology», *Journal of Medical Ethics*, 48, 2022, pp. 278-284.

[120] In this regard, Floridi coined the term «Onlife». Luciano Floridi, *The onlife manifesto: Being human in a hyperconnected era*, Springer, 2015.

[121] Brendan Kelly *et al.*, «Cybersecurity in Healthcare», in Houneida Sakly *et al.* (ed. by), *Trends of artificial intelligence and big data for e-health*, Springer, 2022, pp. 213-231.

[122] Jagpreet Kaur, and K.R. Ramkumar, «The recent trends in cybersecurity: a review», *Journal of King Saud University—Computer and Information Sciences*, 34(8) part B, 2021, pp. 5766-5781.

ed debating about the possibility of recognizing a new right to cybersecurity.[123] In the field of healthcare, ensuring an adequate level of cybersecurity
is essential not only to safeguard and protect the confidentiality of the data
being processed by the systems, but also to guarantee that the standard of
care can be provided to patients and thus to safeguard their right to health.
However, healthcare has so far adopted only «immature cybersecurity practices» even though it is «a prime target for data theft».[124] As a consequence,
and precisely because of the progressive digitization of the field at its various
levels, especially with AI systems, solid cybersecurity programs and practices, based on an adequate set of norms, are all the more necessary.

Finally, questions arise as to whether it should be possible, and, if so, how,
to protect AI systems under the umbrella of Intellectual Property protection
(IP). Various types of protection may be considered: patents, copyright, and
the protection awarded to trade secrets, all of which need to be adapted
to the specificities of both AI and the healthcare sector. The importance
of regulating IP protection for AI systems rests in the necessity to provide
companies and providers who invested in the development of the technology
itself with sufficient rights, especially related to the economic exploitation of
the system, to encourage them to invest in the sector and thus foster technological development.

2.6 Conclusions

The tremendous advantages of the introduction of AI in the various aspects
of healthcare bring along fundamental issues to be addressed both from an
ethical and legal perspective to guide its development through time toward
the implementation of a technology that is trustworthy and respectful of the
fundamental rights of the people involved and values of society.

The list of ethical and legal issues to be addressed in the process, with
no claim to be comprehensive, aimed at providing an overview of those that
are most discussed and generally shared throughout the various jurisdictions and at the international level. Undoubtedly, considering the need to
maintain a context-dependent interpretation of both issues, variations may
occur among countries and according to national perspectives. To attempt to
complete the picture, the following chapter is devoted to providing a similar
overview of the policy issues to be addressed on the matter.

---

[123] Pier Giorgio Chiara, «Towards a right to cybersecurity in EU law? The challenges ahead», *Computer Law and Security Review*, 53, 2024, pp. 1-9.
[124] Brendan Kelly *et al.*, *op. cit.*

3 Policy Challenges in Regulating AI in Healthcare

3.1 A global race to better regulate (or not regulate) artificial intelligence in healthcare

The deployment of artificial intelligence technologies in the biopharmaceutical sector promises to enhance innovation and accelerate the development and delivery of effective treatments for patients. Economically, the projected impact of AI on the global economy by 2030 is considerable, with estimates of 14% of global gross domestic product, half of which derives from productivity gains.[1] Moreover, healthcare was the sector that received the most AI-related investments in 2022, amounting to US $6.05 billion.[2] At the same time, AI in public health holds huge potential, as it is crucial for disease surveillance, early outbreak detection, epidemiological analysis, resource optimization, and patient care improvement, thereby enhancing decision-making and enabling more effective interventions. However, the advancement of AI technology also poses new policy challenges, ethical dilemmas, and safety risks, such as the need for global harmonization of definitions and risk frameworks, the trade-off between regulation and innovation, the optimization of regulatory data processes for validating new AI-based technologies (especially those involving dynamic models or black boxes), the loss of human control and accountability, the issues of transparency and explainability, AI inventorship, the need for AI literacy, and problems of eco-sustainability.

Various governments around the world have initiated legislative, regulatory, and policy initiatives to determine proper use of AI. Furthermore, given that AI use in the healthcare sector presents many distinctive features, especially because it raises questions about fundamental rights (life and health) and is closely related to patient safety, it becomes extremely important to understand how to ensure an ethical use of AI in this specific sector. Thus, a number of governments worldwide are endeavoring to understand how to ensure the responsible use of AI, including in healthcare: In the Executive

---

[1] PwC, *Sizing the prize: what's the real value of AI for your business and how can you capitalize?*, 2017.
[2] Stanford Institute for Human-Centered Artificial Intelligence, *Artificial Intelligence Index Report 2023*, 2023.

Order on AI that US President Joe Biden signed in October 2023, several organizations, including the US Department of Health and Human Services and the US Department of Veterans Affairs were tasked with investigating how to safely implement AI in healthcare; in the United Kingdom, the National Health Service has allocated more than £123 million ($153 million) to the development and evaluation of AI, and a further £21 million to its deployment; similarly, in June, the European Union allocated €60 million ($65 million) to research into AI in healthcare and its deployment.[3]

However, in the meantime, the private sector is also developing its own set of rules: Google has committed to seven AI principles and promised to not deploy AI in four key areas,[4] while Microsoft runs several initiatives committed to the advancement of AI driven by ethical principles.[5] Also, Google and Microsoft, along with DeepMind, Amazon, and IBM, founded Partnership on AI, a not-for-profit coalition that works to «advance responsible governance and best practices in AI,» while Microsoft, Google, OpenAI, and Anthropic founded an industry body (Frontier Model Forum) with the aim to promote «safe and responsible development of frontier AI systems.» On the other side of the Ocean, instead, the European Commission promoted the «AI pact,» aimed at helping stakeholders, including companies, non-for-profit organizations, academics, and civil servants, to prepare for the implementation of the AI Act through the sharing of best practices, internal policies, and voluntary commitments.

Thus, while governments race to regulate AI and companies move to self-regulate their use of AI, this chapter aims to provide an overview of the main policy challenges that countries and the healthcare sector will have to face when it comes to ensuring the responsible use of AI.

---

[3] Augustin Toma *et al.*, «To safely deploy generative AI in healthcare, models must be open source», *Nature*, 624, 2023, pp. 36-38.

[4] Google AI, *Our Principles*, https://ai.google/responsibility/principles/, consulted on November 13, 2024.

[5] Microsoft AI, *Empowering responsible AI practices*, https://www.microsoft.com/en-us/ai/responsible-ai, consulted on November 13, 2024.

3.2 The need for global harmonization in terms of definitions and approaches

As new artificial intelligence tools and applications emerge and proliferate around the world, the demand for a coherent approach becomes more urgent, especially as the aspiration to regulate this technology grows. Indeed, using the same technical terms, agreeing on the same definitions, and adopting the same standards is essential to ensure clarity and consistency, while improving communication and collaboration. Divergence, on the other hand, could lead to cumbersome complexity in the interpretation and implementation of regional and national legislation, as well as legal uncertainty. Also, it is likely that a lot of AI uses will likely span multiple countries, as this technology will break down boundaries. Therefore, the local operating environments need to be as harmonized as possible to promote cross-border leverage of AI opportunities. The benefits of a coherent approach would also include the reduction of costs and the enhancement of international research and trade, interoperability across jurisdictions, confidence in the system by AI users, fairness, and safety. While an extensive overview of the issue has already been provided in Chapter 1, when the challenges related to identify and timely fix the definition of the terms most frequently used on the matter, we believe it is nonetheless of interest here to address the topic from a policy point of view.

In general, it is important that the local efforts to regulate and discuss policies around AI are consistent with the global ones to achieve a common terminology, to avoid creating a landscape characterized by an inconsistent use of definitions and terminology. Now, the challenge is to identify some common terms and definitions to be used by stakeholders around the world.

An emblematic example of the importance of harmonizing terms and meanings is represented by the trajectory of the very same definition of artificial intelligence. In this sense, a first attempt to find a global consensus on a definition of AI to be used as a global benchmark was undertaken by the OECD, which inserted a first early definition in the OECD AI Principles of 2019[6] that are now recognized as the first intergovernmental standard on AI. However, in its final provisions, the Council instructed the Committee for Digital Economic Policy to «monitor, in consultation with other relevant Committees, the implementation of this Recommendation and report to the Council no later than five years following its adoption and regularly thereafter.» Accordingly, in November 2023, OECD member countries had

---

[6] OECD, *AI Principles*, 2019.

to approve a revised version of the organization's definition of an AI system. The updated definition was:

An AI system is a machine-based system that for a given set of human-defined *explicit or implicit* objectives, *infers, from the input it receives, how to generate outputs such as* makes predictions, *content*, recommendations, or decisions *that can* influence *physical* real or virtual environments. *Different* AI systems are designed to operate with varying *in their* levels of autonomy *and adaptiveness after deployment.*

A timely effort was made to enable the European Union to incorporate the OECD definition into the Artificial Intelligence Act, addressing issues related to the EU's desire to maintain semantic alignment with international partners. However, the discussion had to be postponed for several months until the OECD officially confirmed the new definition. Nevertheless, this definition was still subject to intense negotiations throughout the whole process.

However, the quest for a global taxonomy does not end within the European Union. Following the suggestions for concrete activities aimed at aligning EU and US risk-based approaches published in December 2022[7] in the EU–US TTC Joint Roadmap for Trustworthy AI and Risk Management, a group of experts engaged to prepare an initial draft AI terminologies and taxonomies. By May 2023, a total of 65 terms were identified with reference to key documents from the EU and the US.[8] This work advances the EU and the US shared interest to support international standardization efforts and promote trustworthy AI, while complementing the G7 Hiroshima AI process.

In fact, assuming a consensus can be found on the definition of AI, we can delve into even more advanced and specific definitions. For example, what does «trustworthy AI» mean? Over the years, many AI principles trying to describe trustworthy AI features have been developed by policy, regulatory, and professional organizations, including the World Health Organization (WHO)[9] and UNESCO.[10] Also, the US Food and Drug Administration (FDA), Health Canada, and the United Kingdom's Medicines and Healthcare products Regulatory Agency (MHRA)[11] have jointly iden-

---

[7] European Commission, *E.U.-U.S. TTC Joint Roadmap for Trustworthy AI and Risk Management*, 2022.

[8] European Commission, *E.U.-U.S. Terminology and Taxonomy for Artificial Intelligence*, 2023.

[9] World Health Organization, *Ethics and governance of artificial intelligence for health*, 2021.

[10] UNESCO, *Recommendation on the Ethics of Artificial Intelligence*, UNESCO, 2022.

[11] U.S. Food and Drug Administration (FDA), Health Canada, and the United

tified 10 guiding principles that can inform the development of Good Machine Learning Practice (GMLP). Fortunately, there is considerable synergy between these proposed AI principles, which often focus on or refer to the same requirements (such as non-maleficence, fairness, transparency, explainability, and intelligibility, responsibility and accountability, inclusiveness, and equity). This seems to demonstrate that there could be an evolving global consensus on what constitutes responsible AI or trustworthy AI, at least in general terms.

However, a significant challenge presents itself when attempting to categorize the varying degrees of risk related to AI technologies. Indeed, as the risk-based method gains popularity worldwide, with implementation in numerous countries and regions, there is a considerable discrepancy in how unacceptable or high-risk AI systems are defined, leading to an array of more or less extensive classifications. For instance, the European Union has identified certain applications of AI that, due to the recognized potential threat they pose to citizens' rights and democracy, are prohibited by the co-legislators, such as for instance biometric categorization systems that use sensitive characteristics (e.g., political, religious, philosophical beliefs, sexual orientation, race) or indiscriminate scraping of facial images from the internet or CCTV footage to create facial recognition databases. Moreover, a number of AI systems defined in the EU AI Act, which could potentially create an adverse impact on people's safety or their fundamental rights (as protected by the EU Charter of Fundamental Rights), are considered to be high-risk and thus are subject to more obligations. These categories include, for example, access to essential private and public services and benefits (e.g., healthcare), creditworthiness evaluation of natural persons, and risk assessment and pricing in relation to life and health insurance. Therefore, while a more extensive discussion of the approach adopted by the European Union is provided in Chapter 4, it may be already underlined here that within the European Union, the risk classification is based on the intended purpose of the AI system, in line with the existing EU product safety legislation. It means that the risk level depends on the function performed by the AI system and on the specific purpose and modalities for which the system is used.

In the United States, on the other hand, there is only limited guidance from the federal government agencies and Congress on which AI use cases are high risk, as a unified definition is still lacking. For example, the National Institute of Standards and Technology's (NIST's) Artificial Intelligence

Kingdom's Medicines and Healthcare products Regulatory Agency (MHRA), *Good Machine Learning Practice for Medical Device Development: Guiding Principles*, 2021.

Risk Management Framework (AI RMF),[12] which is a voluntary consensus framework that builds on the Organization for Economic Cooperation and Development's Framework for the Classification of AI Systems, is intended to support AI stakeholders in self-regulation by mapping, measuring, and managing AI risks and building AI risk management governance programs. However, it does not explicitly define high-risk AI use cases, similar as the White House Office of Science and Technology Policy's Blueprint for an AI Bill of Rights,[13] Government Accountability Office's non-binding AI accountability framework,[14] and President Biden's Executive Order on the Safe, Secure, and Trustworthy Development and Use of Artificial Intelligence (which is actually based on the AI RMF)[15], only provide general insights into the risks and benefits that various AI technologies can produce.

The differences between the US and EU approaches were also highlighted at the G7 Summit held in May 2023, in which, while the focus was on the emerging need for trustworthy AI along with the necessity for interoperability among AI governance frameworks, it was also stressed the need to align different risk-based frameworks as designed in various environments. In fact, specifically, a comparison between the EU AI Act's categorization and the US NIST AI Risk Management Framework was conducted, revealing significant differences in risk assessment.[16] This can be explained by the fact that the US and the European Union have distinct goals when it comes to AI regulation, which are reflected in their differing approaches. In fact, while the latter is worried about the possible effects of strict regulation on innovation and competition, the former prioritizes the protection of citizens' fundamental rights. Anyway, this global framework also implies that, once a consensus is reached on adopting a risk-based approach, each government can still exercise some discretion in defining the acceptable and unacceptable risks within their jurisdictions, based on their local values and needs.[17]

---

[12] National Institute of Standards and Technology, *AI Risk Management Framework: Initial Draft*, 2022; National Institute of Standards and Technology, *AI Risk Management Framework*, 2023.

[13] The White House, *Blueprint for an AI Bill of Rights*, 2022.

[14] U.S. Government Accountability Office, *Artificial Intelligence: An Accountability Framework for Federal Agencies and Other Entities*, 2021.

[15] The White House, *Executive Order on the Safe, Secure, and Trustworthy Development and Use of Artificial Intelligence*, 2023.

[16] Christina Todorova, *et al.*, *The European AI Tango: Balancing Regulation, Innovation and Competitiveness*, in Human Centered AI Education and Practice Conference, December 14 & 15, 2023, Dublin, Ireland. ACM, New York, NY, USA.

[17] Edoardo Chiti, and Barbara Marchetti, «Divergenti? Le strategie di Unione europea e Stati Uniti in materia di intelligenza artificiale», *Rivista della Regolazione dei Mercati*, Fascicolo I, 2020.

3.3 Balancing limitations to ensure responsible use of AI with the need to avoid the inhibition of innovation. California as a case study and European Union's strategy to enhance competitiveness

One of the most formidable challenges for policymakers is to strike a balance between ensuring the responsible use of AI and avoiding the inhibition of innovation.[18] The challenge lies not just in deterring domestic companies from moving to jurisdictions with more advantageous conditions but also in drawing fresh investments and promoting innovation and competitiveness via an accommodating regulatory framework. Indeed, companies are always seeking the optimal conditions for their business, and the role of the regulatory framework in influencing their decisions should not be underestimated.[19]

To appreciate the perspective of companies, it is instructive to examine the policy positions of Technet[20] (a trade group that comprises various companies such as Apple, Google and Amazon), which articulate what its members would and would not support in terms of AI regulation, especially in the United States, where most of them are based and/or were born. For instance, these positions clearly state that policymakers should refrain from imposing «blanket prohibitions on artificial intelligence, machine learning, or other forms of automated decision-making» and «reserve any restrictions only for specific, identified use-cases that present a clearly demonstrated risk of unacceptable harm, and narrowly tailor those requirements to the harms identified.» Moreover, «federal regulations should not force companies to provide proprietary or protected information. Enforcement should be limited to the relevant agencies and avoid private rights of action.» The objective of companies is, understandably, to safeguard their data and IP, while avoiding onerous legislation that would impede their efforts to develop new technologies.

The discussion surrounding the regulation of AI, while still fostering innovation, is active on both sides of the Atlantic. In the US, an illustrative case of the issue is the shaping of the policy and legal framework of California, which hosts Silicon Valley and constitutes one of the most innovative environments in the world.[21] Also, it is worth noting that in the US companies

---

[18] Gabriele Maglio, «*Bilanciare Regolamentazione e Innovazione: Le Sfide del Mercato Digitale Europeo*», Luiss Policy Observatory, Policy Brief n. 2/2024.

[19] *Ibid.*

[20] TechNet, *2023 Federal Policy Principles*, 2023; and TechNet, *2024 State Policy Principles*, 2024.

[21] Executive Department of California, Executive Order N-12-23, 2023.

have a significant influence on politics, as they provide substantial funding for electoral campaigns: For instance, during the 2020 cycle, Big Tech companies spent $124 million in lobbying and campaign contributions, breaking their own records from past election cycles. Amazon's spending increased by 30%, and Facebook's spending surged by an astonishing 56%.[22]

The Governor of California issued the Executive Order (E.O.) N-12-23 in September 2023, after several bills in California had been obstructed by the lobbying efforts of major companies. Also, many subsequent attempts in 2024 to regulate AI failed, with California Governor Gavin Newsom vetoing the artificial intelligence safety bill known as SB 1047, officially because the legislation didn't take into account whether an AI system was deployed in high-risk environments, involved critical decision-making or the use of sensitive data. Instead, according to the Governor, the bill applied stringent standards to even the most basic functions, so long as a large system deploys it. In this particular instance, California's debate gained significant attention because the proposed bill would have effectively established a national standard for the technology.[23] Thus, Governor Newsom encountered intense lobbying from major economic and political figures in the state, such as Hollywood actors, leading tech companies and investors, and powerful House members including former Speaker Nancy Pelosi. While some leading researchers, along with Elon Musk (CEO of Tesla, SpaceX and Twitter and co-founder of Neuralink and OpenAI), backed the measure to help reduce possible risks to the public, other opponents like Google and OpenAI contended that its requirements would place excessive burdens on developers, particularly smaller startups.

In any case, California's Executive Order that was approved in 2023 acknowledged the exceptional environment of California state, which is «leading the world in GenAI innovation and research and it is home to 35 of the top 50 AI companies in the world,» and declared that «the State of California seeks to realize the potential benefits of GenAI for the good of all California residents, through the development and deployment of GenAI tools that improve the equitable and timely delivery of services, while balancing the benefits and risks of these new technologies.» To achieve these objectives, the EO included provisions to direct state agencies and departments to

---

[22] Public Citizen, *Big Tech, Big Cash: Washington's New Power Players*, 2021.
[23] California state laws often established a national standard, especially when related to technology and in absence of an effective federal regulation. For example, California's privacy law served as a basis for developing multiple states' privacy laws after it was enacted. This aspect is better discussed in Chapter 5, devoted to the United States regulatory framework.

perform a joint risk analysis of potential threats to and vulnerabilities of California's critical energy infrastructure by the use of GenAI, while it required state agencies to issue general guidelines for public sector procurement, uses, and required training for the application of GenAI, also building on the White House's Blueprint for an AI Bill of Rights and the National Institute for Science and Technology's AI Risk Management Framework. This indicated that California was already attentive to the global scenario and that it endorsed the same risk-based approach that had been adopted by the European Union and the NIST as the global standard.

Interestingly, to further foster innovation, the state decided to establish the infrastructure needed to conduct pilots of GenAI projects, including California Department of Technology-approved environments or sandboxes to test such projects. Regarding this point, it should be highlighted that AI sandboxes are one of the most popular approaches taken into consideration as a viable option to promote testing out new innovation without the heaviness of policies/regulations that dictate it. Then, once a concept is tested out and proved, the governance body can decide if it has met sufficient safety/ethics protocols to allow it to leave the sandbox and become a real product. This approach had already been chosen by countries such as Singapore[24] and the United Kingdom[25], as well as by the European Union Artificial Intelligence Act, which promotes regulatory sandboxes and real-world testing, established by national authorities to develop and train innovative AI before placement on the market.[26]

Lastly, it has to be noted that the Executive Order specifically provides for the «Governor's Office of Business and Economic Development, in consultation with the Government Operations Agency, to be directed to pursue a formal partnership with the University of California, Berkeley, College of Computing, Data Science, and Society and Stanford University's Institute for Human-Centered Artificial Intelligence to consider and evaluate the impacts of GenAI on California and what efforts the State should undertake to advance its leadership in this industry. As part of this effort, beginning in

---

[24] Infocomm Media Development Authority, *First of its kind Generative AI Evaluation Sandbox for Trusted AI by AI Verify Foundation and IMDA*, press release, October 31, 2023.

[25] The Information Commissioner's Office (ICO) defines the regulatory sandboxes as: «a free service developed by the ICO, to support organisations who are creating products and services which utilize personal data in innovative and safe ways.,», consulted on November 13, 2024.

[26] Thomas Buocz, Sebastian Pfotenhauer, and Iris Eisenberger, «Regulatory Sandboxes in the AI Act: Reconciling Innovation and Safety?», *Law, Innovation and Technology* 15 (2), 2023, pp. 357–89.

the fall of 2023, those agencies are directed to work with the University of California, Berkeley, College of Computing, Data Science, and Society and Stanford University's Institute for Human-Centered Artificial Intelligence to develop and host a joint California-specific summit in 2024, to engage in meaningful discussions and thought partnership about the impacts of GenAI on California and its workforce and how all stakeholders can support growth in a manner that safeguards Californians.» Grasping this key aspect is crucial for comprehending California's approach to regulating AI: active legislative collaboration involving all the pertinent parties is critical for establishing an efficacious legal structure for artificial intelligence. As innovation predominantly occurs in universities or university-affiliated projects, there must be an open dialogue between them and private corporations. Given that technology companies are synonymous with California, it is imperative for the state to maintain strong ties without imposing undue strain, ensuring that innovation continues to receive robust support and stimulation.

The European Union is confronted with the same dilemma of balancing innovation and protection of rights, but with very different premises and a different outcome.[27] The efforts of Asia and the United States to become AI leaders seem to have relegated the European Union to a secondary position, which is now seeking to promote innovation within its borders and to attract new investments, in order to create a vibrant AI environment. AI regulation plays a pivotal role in fostering the European hub for innovation, but it also represents a major challenge, as a too-prescriptive approach may inhibit that same innovation and be counterproductive to achieving those goals.[28]

The report on the future of European competitiveness, presented by former European Central Bank governor Mario Draghi in September 2024,[29] clearly framed the issue. The regulatory framework in Europe may create harm to small businesses or deter large companies: Frequent changes and the complexity of regulations often create high compliance costs for small- and medium-sized Enterprises (SMEs), while discouraging investments from large companies.[30] Additionally, medical companies may be partic-

---

[27] KaterinaYordanova , «The EU AI Act—Balancing Human Rights and Innovation Through Regulatory Sandboxes and Standardization», *TechREG* Chronicle, 2022.

[28] Also, on the Act's potential impact on developing countries and concerns that the Act's uniform standards could potentially exacerbate the digital divide and create barriers in global AI innovation and trade, see Qiang, Ren., and Jing Du, «Harmonizing innovation and regulation: the EU artificial intelligence act in the international trade context», *Computer Law & Security Review*, 54(106028), 2024.

[29] Mario Draghi, *The future of European competitiveness*, 2024.

[30] *Ibid.*

ularly affected by restrictive regulations like GDPR, which include severe limitations on the secondary use of data for research, hindering innovation, and significantly restricting the ability to conduct clinical trials and develop new treatments.[31] On one side, the report emphasizes the importance of integrating AI vertically into European industry to unlock higher productivity and protect Europe's social model. On the other hand, Draghi's report recognizes that while the ambitions of the EU's GDPR and AI Act are admirable, their complexity and potential for inconsistencies may hinder AI development by EU industries. In practical terms, differences in how Member States implement and enforce GDPR, along with overlaps with the AI Act, may create regulatory uncertainty and higher barriers for EU researchers to innovate. Altogether, this places European companies at risk of missing early AI advancements. To mitigate this risk, the report recommends simplifying rules and harmonizing GDPR implementation across Member States, while eliminating overlaps with the AI Act to avoid penalizing EU companies in AI development and adoption.

Furthermore, many criticisms of the European approach have been voiced by the Big Tech US companies, as they advocate for more flexible frameworks like the ones in their homeland. For instance, one of the AI leaders, Sam Altman of OpenAI, expressed concerns about the European Union's AI regulations, suggesting that the burden of compliance could lead to ceasing operations in Europe.[32] At the same time, Meta announced in summer 2024 that it won't be launching its upcoming multimodal AI model (capable of handling video, audio, images, and text) in the European Union, citing regulatory concerns; the decision prevented European companies from using the multimodal model, despite it being released under an open license.[33] Likewise, and within a similar timeframe, Apple announced that it would postpone the European release of its generative AI software suite, primarily due to regulatory ambiguities and the impact of the European Union's Digital Markets Act.[34]

However, the specificities of the European environment need to be taken into account. In fact, the EU's business ecosystem is predominantly characterized by SMEs, which represent the majority of the business landscape and are the real driving forces behind AI innovation in Europe, counting

---

[31] *Ibid.*

[32] Richard Waters, Javier Espinoza, and Madhumita Murgia, «OpenAI warns over split with Europe as regulation advances», *Financial Times*, May 25, 2023.

[33] Cynthia Kroet, «Meta stops EU roll-out of AI model due to regulatory concerns», *Euronews*, July 18, 2024.

[34] Foo Yun Chee, «Apple to delay launch of AI-powered features in Europe, blames EU tech rules», *Reuters*, June 21, 2024.

for 99% of all businesses in the EU.[35] However, as these enterprises pursue the path of innovation, they encounter significant regulatory obstacles, with one of the major challenges posed by the AI Liability regulations,[36] which have the potential to significantly increase the regulatory burden for smaller firms. An immediate consequence is a rise in the cost of insurance against AI Liability, which may act as a deterrent for many businesses, particularly startups, from venturing into high-risk AI domains.[37] In fact, the relationship between rapid technological adoption and adherence to regulatory standards presents challenges, particularly for startups that might prioritize technical solutions over regulatory considerations, leading to costly missteps;[38] indeed, unlike their larger counterparts, startups often deal with limited resources and expertise when facing AI regulations.[39] To close the understanding and compliance divide, agencies supported by the government might begin offering advisory services on AI regulations or developing regulatory testing environments, with the aim of supporting startups in managing AI regulations adeptly.

Finally, within the healthcare industry, significant apprehension regarding the EU AI regulatory landscape still persists. For example, the European Federation of Pharmaceutical Industries and Associations (EFPIA), which advocates for the pharmaceutical sector in Europe, shared its reservations about the Artificial Intelligence Act regarding the classification of all AI-driven medical devices as high-risk. They are also concerned about the interplay with the current European Union Medical Device Regulation, which may have already established similar overlapping standards affecting all medical devices, asking for further guidance, clarity, and harmonization between all the existing regulations.[40]

---

[35] Small- and medium-sized enterprises data and definition are provided by the European Commission at *SME definition - European Commission* at https://single-market-economy.ec.europa.eu/smes/sme-fundamentals/sme-definition_en, as well as they are defined by the defined in the EU recommendation 2003/361 and the *The revised user guide to the SME definition*, 2020. Also, see the Eurostat, Report *EU Small and Medium-Sized Enterprises: An Overview*, 2022.

[36] European Commission, Liability Rules for Artificial Intelligence, 2022.

[37] Benjamin Mueller, *How Much Will the Artificial Intelligence Act Cost Europe?*, Center for Data Innovation, July 2021.

[38] *Ibid.*

[39] Pierre Crespi, *European AI Act—How will this regulation affect SMEs?*, European Digital SME Alliance's, July 6, 2021.

[40] Digital Health Working Group, *EFPIA's Position Paper on Artificial Intelligence*, 2020.

3.4 Considering healthcare sector-specific needs when developing AI regulations

Regulatory frameworks for AI may need to take into account the specificities of the biopharmaceutical and healthcare sector, taking into serious account the ethical and legal-theoretical challenges highlighted above. For example, a generic and uniform approach across sectors could unintentionally hamper innovation that leads to faster and better treatments for patients, especially if there are no specific exemptions or exceptions for AI in research in a biopharmaceutical context. As the healthcare sector is inherently complex and diverse, and covers a wide range of applications, including medical diagnosis, pharmaceutical research, patient monitoring, personalized therapies, and more, each of these areas may pose different challenges and risks that may require a tailored legislative/regulatory response.

Coherently with the ethical framework discussed in Chapter 2, one reason for adopting sector-specific provisions is the high sensitivity of health data and the imperative of ensuring patient privacy in the healthcare industry. A case in point is the issue of genomic data (data containing information on the base sequence in an individual's genome, including detailed and sensitive information about an individual's genetic makeup) which are so intrinsically personal that their anonymization presents serious challenges and under certain circumstances may not be possible at all, while still retaining very sensitive information, such as the individual's predisposition to develop a condition before the onset of symptoms.[41] In this case, sector-specific approaches could establish specific requirements to ensure that AI applications in the healthcare sector are safe and effective while adequately protecting patients' data.

Another reason for considering sector-specific provisions is the dynamic and evolving nature of AI in healthcare, with new applications and discoveries emerging regularly. Sector-specific approaches could allow for more flexibility and adaptability in regulating new technologies, based on the level of risk they entail, without having to revise the entire regulatory framework.

Clear and well-defined rules for the use of AI in healthcare would benefit multiple stakeholders: First, they would incentivize the private sector to invest in AI in medical practices, by reducing the investment risk associated with regulatory uncertainty; second, they would clarify the roles and responsibilities of technology providers and medical professionals, which is

---

[41] Gemma Bilkey *et al.*, «Genomic testing for human health and disease across the life cycle: applications and ethical, legal, and social challenges», in *Frontiers in Public Health*, 7(40), 2019.

crucial in the context of human health in determining the liability in case
of errors or harm caused by AI; third, sector-specific approaches would also
foster public trust. As AI in healthcare involves health and life-related issues, it is essential that the public has confidence in the technologies used.

3.5 The risk-based approach applied to the healthcare sector: cases of use

Before being integrated into clinical practice, AI technologies must undergo rigorous evaluation.[42] For example, the first AI-based device to receive
market authorization from the FDA was subjected to an extensive prospective comparative clinical trial involving 900 patients across diverse sites.[43]
Stringent regulatory standards must govern the approval of AI technologies
as medical devices, mainly because the decision support provided is optimized and personalized continuously in real time, and the performance of
AI depends strongly on the training datasets used,[44] resulting in a large risk
of AI performing less well in real practice or on another group of patients
or institutions.[45] Therefore, a thorough assessment of AI's performance and
safety is imperative before its incorporation into routine clinical practices.

In January 2025, the US Food and Drug Administration released a draft
guidance on AI in drug development[46] that strongly supports the European
Union AI Act's fundamental principles, as both frameworks place a high
priority on using a risk-based approach. In fact, the FDA highlights the
necessity of adjusting regulatory monitoring to the specific risks connected
to various AI uses, in a similar way to the European AI Act, which classifies
AI systems according to their degree of danger, with stricter regulations
applied to higher-risk systems. Also, both the AI Act and FDA frameworks
stress how crucial model credibility and usage context are: On one side,

---

[42] FDA, Artificial Intelligence and Machine Learning in Software as a Medical
Device, 2024.

[43] Michael D Abràmoff *et al.*, «Pivotal trial of an autonomous AI-based diagnostic
system for detection of diabetic retinopathy in primary care offices», *NPJ Digital
Medicine*, 1(39), 2018.

[44] Junhua Ding, and Xinchuan Li, *An approach for validating quality of datasets for
machine learning*, 2018 IEEE International Conference on Big Data (Big Data), 2018.

[45] Dong Wook Kim *et al.*, «Design Characteristics of Studies Reporting the Performance of Artificial Intelligence Algorithms for Diagnostic Analysis of Medical
Images: Results from Recently Published Papers», *Korean Journal of Radiology*, 20(3),
2019, pp. 405-410.

[46] FDA, «Considerations for the Use of Artificial Intelligence to Support Regulatory Decision-Making for Drug and Biological Products», *Draft Guidance for Industry
and Other Interested Parties*, 2025.

the FDA emphasizes that confidence in the AI model's performance in a particular setting is essential for the proper use of AI in drug development, on the other, the EU AI Act highlights the importance of making sure AI systems are trustworthy and dependable for the purposes for which they are designed.

The risk-based approach is often regarded as the most appropriate one to safeguard patients without impeding innovation. However, its objectives can be achieved only if it is implemented effectively, without imposing excessive requirements on those AI systems that only present limited risk and are thus not likely to pose a significant risk of harm to the health and safety of individuals and society. Therefore, when the risk-based approach is applied, the level of transparency, explainability, and communication required by AI legislation or regulation should always be proportionate to the level of risk. This implies that, in the healthcare sector, each risk category should be identified and accompanied with specific, clear, and proportionate regulatory requirements based on different levels of risk, intended use, and context. For example, in drug development, the phase of research and development in which the AI system is used would constitute a crucial factor in risk assessment. For instance, most uses of AI in drug discovery are related to optimizing drug designs, selecting the right target in the body, and predicting new small molecules with desirable properties. For this reason, they are likely to fall into a low-risk category, even if there should be room for flexibility to allow the categorization of some uses in this setting as high-risk should unusual circumstances apply. Similarly, many use cases of AI in clinical trial settings will likely pose a high risk, but it would be overly restrictive to categorize all uses of AI in this setting as high risk without allowing for flexibility in case circumstances differ. Conversely, in the hospital context, for example, there may be AI systems that are used in an «indirect way» to enhance healthcare efficiency, such as to complete administrative tasks, create reports, and help in general the hospital organization, thus presenting a low risk; other AI systems, instead, can be used in a «direct way,» supporting the clinical decision of the healthcare professional, evaluating a patient case, or providing a medical diagnosis, thus presenting a higher risk.

It is important to remember that artificial intelligence serves as an essential tool in multiple ways, with applications in both the public and the private sector. First, AI is crucial in policymaking for its ability to process extensive healthcare data and generate valuable insights. However, as mentioned, the significant risk of bias in datasets can lead to unequal and potentially harmful outcomes; hence, a risk-based approach is essential to ensure that AI applications are both effective and equitable, mitigating potential biases and safeguarding public trust. At the same time, also the employment of AI in the healthcare industry faces numerous challenges, equally matched by its potential. AI and machine learning are being effectively leveraged in various stages of the drug lifecycle, including development, discovery, and conducting clinical trials. The utility of AI extends even further, encompassing activities such as medical diagnosis, customization of treatment plans, correlation analysis between clinical data and patient outcomes, as well as uncovering novel connections within genetic codes. AI supports patients across their healthcare journey, aids with monitoring systems, or even helps healthcare professionals manage routine administrative duties up to enhancing the capabilities of surgery-assisting robots. Therefore, it is essential to thoughtfully evaluate the associated risks and benefits of these implementations to safeguard the citizens' right to health effectively.

3.6 Regulating dynamic AI models and black boxes

The application of machine learning algorithms and fundamental models is rapidly expanding within the realm of digital healthcare. Although medical devices are subject to stringent regulations, the integration of AI/ML presents novel challenges.[47]

Initially, with the surge of digital health devices hitting the market, we may see an increase in the number of borderline devices, meaning those with a regulatory status that isn't clearly defined. This can happen because many of them could be classified as lifestyle or general well-being products. Additionally, there's a divergence in the way digital health devices are regulated in the US compared to the EU. The FDA uses a risk-based discretionary approach to determine regulatory oversight, differentiating between devices that require regulation and those that don't based on the level of risk. On the other hand, the EU employs a device qualification criterion without such discretionary power. Under the MDR/IVDR, a device is subject to regulation if it meets the qualification requirements.

Nonetheless, with the enforcement of the Artificial Intelligence Act and the EU's shift toward a risk-based framework, there exists a potential for the regulatory systems of the US and EU to align more closely, despite their different regulatory techniques. As it stands, risk assessment already influences the differentiation of medical devices from wellness-related devices within the EU context.

---

[47] Johan Ordish, Hannah Murfet, and Alison Hall, *Algorithms as medical devices, PHG Foundation*, 2019.

Machine learning in the field of medicine serves multiple purposes, including automation or assistance in tasks typically performed by human professionals, such as segmenting medical imagery. It's also utilized in analyzing large datasets to uncover new patterns and insights that may lead to the identification of new diseases or the discovery of biomarkers for potential drug targets. Furthermore, machine learning is essential in the predictive analysis of health and disease, utilizing complex pattern recognition for uses such as detecting illnesses, providing diagnostics, and aiding in clinical decision-making processes.[48] This implies that the volume of data and the degree of human interaction differ significantly based on the machine learning method employed. Consequently, the demands for a machine learning algorithm can be greatly influenced by the specific context.[49] Furthermore, the regulatory approval process for new digital technologies is inherently complex and often demands additional testing data or repeated tests.

It's important to recognize that many AI applications incorporate learning algorithms that enable them to adapt and modify over time. Consequently, these systems accumulate training data progressively, and the model undergoes continuous incremental learning since machine learning algorithms operate on datasets. Nonetheless, it is not essential for all AI/ML technologies to undergo incessant retraining, particularly in the field of healthcare. For instance, clinical decision support systems (CDSSs) provide physicians with evidence-based recommendations at the point of care that enhance clinical efficiency; however, rule-based CDSS can remain effective with periodic updates based on new medical guidelines or regulatory changes.

Therefore, the regulation of new technologies can be difficult, particularly when they exhibit two characteristics: being «dynamic,» as they evolve with advancements in cybersecurity and continuous learning from ongoing data, making it impractical to rely on current regulatory procedures for continual re-validation; and being «opaque,» where their inner mechanisms are not clear or are understood but not interpretable by humans, which may necessitate sophisticated testing methods for comprehension.

In the context of AI/ML applications, regulators must differentiate between machine learning models that continually learn from streaming data and those that do not update. Models that incorporate new data or retrain could be less stable, as minor modifications in the data or model may lead to vastly divergent outcomes. Furthermore, existing medical device regula-

---

[48] Sobia Raza, *et al.*, *The personalised medicine technology landscape*, PHG Foundation, 2018, pp 96-99.
[49] Andrew L. Beam, and Isaac S. Kohane, «Big data and machine learning in healthcare», *JAMA*, 319(13), 2018, pp. 1317-1318.

tions in European Union, such as MDR/IVDR and harmonized standards, are not designed to oversee machine learning models that are in a state of constant retraining, since they currently only account for changes in devices.

For what concerns opaque machine models, the main issue will be interpretability.[50] This is the case of what is usually defined as a black box algorithm:

> Algorithms can be opaque for multiple reasons. Sometimes, algorithms are non-transparent because, while they may rely on explicit rules, those rules are too complex for us to explicitly understand—for example, patients whose measurements place them in a particular region of n-dimensional (where n is large) characteristic-space are at a higher risk of stroke. In particular, these rules may be impossible to explain or to understand by following the process of scientific/medical discovery: mechanistic lab experiments followed by confirmatory clinical trials. Other times, the relationships used in a black box algorithm are literally unknowable because of the machine-learning techniques employed— that is, no one, not even those who programmed the machine-learning process, knows exactly what factors go into the ultimate decisions. A key distinguishing feature of black-box algorithms, as the term is used here, is that it refers to algorithms that are inherently black box (i.e., their developers cannot share the details of how the algorithm works in practice)—rather than to algorithms that are deliberately black box (i.e., their developers will not share the details of how the algorithm works). Black-box algorithms are especially likely to evolve over time as they incorporate new data into an integrated process of learning-and applying.[51]

Undoubtedly, this might present a challenge when evaluating AI/ML safety or efficacy,[52] despite the absence in the European Union of a specific mandate within the MDR/IVDR or related standards for machine learning to be transparent or subject to white box testing (which entails testing at the code level by checking individual lines of code for accuracy). This implies that black box testing, also referred to as specification-based testing, a method that perceives the functionality of software through the relationship between inputs and outputs and assesses the software's functionality without

---

[50] Paul C. Jorgensen, *Software testing: A craftsman's approach*, CRC Press, 2014, pp 5-9.
[51] W. Nicholson Price II, «Regulating black-box medicine», *Michigan Law Review*, 116, 2017.
[52] Riccardo Guidotti, *et al.*, «A Survey of Methods for Explaining Black Box Models,» *ACM Computer Surveys*, 2019; 51(5), p. 5.

requiring access to its source code, would be consistent with existing regulations.[53]

Taking everything into account, it becomes clear that not every AI/ML technology presents major regulatory challenges. Certain technologies are neither continuously self-improving nor inherently non-transparent. However, when AI/ML systems are either changing dynamically or are not transparent, the current regulatory tools might not sufficiently ensure their safety and efficacy. For these reasons, regulators should start by determining if current medical device oversight regulations are appropriate. Typically, these regulations cover a wide range, from protecting the data used for training models to the requirements for obtaining CE marking and the legalities surrounding liability in the event of malfunctions. Sometimes, the rules are tailored specifically for certain types of digital health devices and can require significant effort from developers. Hence, before considering new regulatory measures, we should first assess and ascertain whether adjustments or modifications are needed within the current regime. Afterwards, alternatives to frequent revalidation and ongoing monitoring may be explored. For instance, regulatory validation may focus on the core technology rather than every new application of it. Once in use, there could be specific standards to decide when changes to the software are substantial enough to need another review, ensuring that no new risks come up. Additionally, there could be requirements for regular audits at set times to check for any risky changes from the original performance.[54] Furthermore, health authorities will need the resources to train and hire professionals who have the needed skills to provide ongoing advice on associated health safety, efficacy, and quality.

3.7 Ensure accountability and liability even where there may be a significant loss of human control

A Pew Research Center poll[55] reveals substantial unease amongst Americans regarding the incorporation of AI into their personal healthcare. In fact, the data shows that 60% of US adults would be uncomfortable if their healthcare professionals depended on artificial intelligence to make diagnostic

---

[53] Paul C. Jorgensen, *op.cit.*, pp. 8-9.

[54] Martin McKee and Olivier J. Wouters, «The challenges of regulating artificial intelligence in healthcare—comment on clinical decision support and new regulatory frameworks for medical devices: are we ready for it?—a viewpoint paper», *International Journal of Health Policy and Management*, 12(2023), p. 7261.

[55] Alec Tyson, *et al.*, *60% of Americans Would Be Uncomfortable With Provider Relying on AI in Their Own healthcare*, Pew Research Center, 2023.

and treatment decisions. Also, a Sogolytics survey[56] conducted in July 2023, polled US adults about their level of ease with AI use across various sectors. Results indicated that AI applications in banking/finance as the ones causing the most discomfort (35%), closely followed by healthcare (at 34%). In fact, when asked to prioritize the advantages of AI, the majority of respondents ranked the improvement in diagnostic accuracy as the least desirable benefit of AI, whereas personalized recommendations were seen as the most beneficial. Furthermore, some concerns regarding the use of AI were also brought to light, with many people particularly worried about diagnostic inaccuracy. This suggests that individuals place more trust in human judgment than AI when it comes to health matters and that they view AI more as an aid rather than a replacement for healthcare professionals, particularly in critical areas like diagnoses.

As already underlined in Chapter 2, populations that have already experienced prejudice in the past frequently maintain mistrust toward the healthcare system and the technologies employed, fearing that AI may perpetuate or worse social and health inequities. These concerns are justified in light of previous events that showed how much AI is likely to incorporate existing human biases and discriminatory behaviors into its algorithms. For example, an AI-enabled pulseoximeter incorrectly read higher blood oxygen levels in patients with darker skin tones, leading to insufficient treatment of their hypoxia.[57]

To be fair, initial research on the accuracy of AI in clinical settings examined the diagnostic performance of deep learning models against that of medical professionals who made diagnoses from medical imaging. These studies indicate that AI can accurately identify diseases in 87% of instances, while the diagnostic accuracy among healthcare professionals is 86%.[58] While further research is considered essential to strengthen the outcomes' validity of this study, it seems evident that in the future AI could match or surpass healthcare professionals in performing certain tasks.[59]

Yet, AI applications in healthcare aren't limited to aspiring to replace humans in complex roles. In fact, AI could also perform routine or administra-

---

[56] «Public Pulse on AI Across Industries: Deep Dive into Patient Comfort, Concerns, and Hopes for AI in Healthcare», *Sogolytics*, 2023.

[57] Michael W. Sjoding *et al.*, «Racial bias in pulse oximetry measurement», *The New England Journal of Medicine*, 383(25), 2020, pp. 2477–2478.

[58] Xiaoxuan Liu, *et al.*, «A comparison of deep learning performance against health-care professionals in detecting diseases from medical imaging: a systematic review and meta-analysis», *The Lancet Digital Health*, 1(6), 2019, pp. e271-e297.

[59] Jeffrey De Fauw *et al.*, «Clinically applicable deep learning for diagnosis and referral in retinal disease», *Nature Medicine*, 24(9), 2018, pp. 1342-1350.

tive tasks, for example through a chatbot that responds to patient inquiries or a gadget that monitors critical health metrics like blood sugar and heart rate, thus reducing the need for doctor visits. Moreover, as technologies like smartwatches become common for personal health tracking, public confidence in AI could grow while amassing more data to train these technologies with greater diversity and reach.

While AI has the potential to be a valuable asset for healthcare providers, concerns about accountability, liability, and safety remain for both medical staff and patients.[60] These have already been addressed in Chapter 2 from an ethical perspective, and in the following pages, they will be discussed from a policy perspective.

Artificial intelligence has the potential to transform the traditional patient–clinician dynamic, which is rooted in trust and empathy. Typically, clinicians discuss options with patients, creating a care plan through mutual decision-making and informed consent, and take into account both patient preferences and medical requirements which are adequately discussed. However, AI is likely to strongly influence this relationship, especially in cases in which clinicians may have limited oversight over AI-generated recommendations and or a general lack of clarity about how AI operates, particularly with so-called «black box» systems. This integration raises questions about the appropriate level of clinician accountability when patients suffer harm as a result of AI involvement in medical decisions,[61] especially because healthcare professionals might struggle to comprehend the reasoning behind the recommendations provided by a non-transparent AI system for patient treatment.[62]

If we believe that artificial intelligence in medicine should be used in a consultative role, and that thus human clinicians should retain responsibility over decisions because they have the final say, there are two possible scenarios to consider: In the first one, clinicians would still need to spend time developing their own assessments, which diminishes the usefulness of the AI system; in the second one, they would be able to rely heavily on the AI's recommendations, but their accountability would be undermined. In cases where clinicians do not actively oversee the AI's advice, and given the lack of transparency of many such systems, the imperative to maintain safety in-

---

[60] Ibrahim Habli *et al.*, «Artificial intelligence in healthcare: accountability and safety», *Bulletin of the World Health Organization,* 98(4), 2020, pp. 251-256.

[61] Helen Smith, «Clinical AI: opacity, accountability, responsibility and liability», *Artificial Intelligence and Society,* 36, 2021, pp. 535-545.

[62] Martin A. Makary and Michael Daniel, «Medical error-the third leading cause of death in the US», *BMJ* (Clinical research ed.), 353(2139), 2016.

creases for both healthcare providers and patients. This assurance of safety is critical in maintaining confidence that patient risk is being reduced as much as is practically feasible.

There's a case to be made for holding clinicians accountable for the AI tools they implement in their practice, especially since they're already required to uphold professional accountability through codes of conduct or national legislation.[63] Since interpreting AI outcomes can be challenging, establishing a robust accountability framework is crucial for maintaining transparency and gaining patient trust.

Yet, a parallel consideration must be made about the obligation of technologists to reveal the origins and varieties of data in the datasets used for training AI, along with its software limitations such as potential data bias. They could also face personal and professional responsibility if their AI systems cause harm to patients when used in clinical environments. Even without direct patient contact, technologists have a duty since they design AI tools that support healthcare professionals in clinical decision-making, a context in which transparency is crucial. When an AI system is set to function autonomously without clinician oversight,[64] it is wise to consider involving AI developers and systems safety engineers in evaluating responsibility for any patient harm and ensuring safety. Furthermore, accountability should also cover data quality in order to avoid datasets that are poor or biased toward higher-income regions, and that would neglect underprivileged groups and produce flawed results with potentially severe consequences. For example, because the algorithms are frequently trained using data largely from white populations, AI dermatological tools have demonstrated a worse diagnosis accuracy for skin lesions and rashes in Black patients compared to white patients.[65] To resolve these challenges, AI developers must consistently ensure ample transparency concerning the algorithm's use and the dataset's quality.

Civil liability is crucial to ensure that individuals impacted by AI-induced harm have recourse for compensation and redress, despite the potential difficulties in pinpointing fault and accountability among those responsible for creating and implementing AI technologies. Of course, the complexity of seeking compensation should not be a barrier for victims as it could erode justice and reduce motivation among stakeholders in the AI sector to

---

[63] European Parliament, *Report with recommendations to the Commission on Civil Law Rules on Robotics—Report—A8-0005/2017*, 2017.

[64] Future Advocacy for the Wellcome Trust, *Ethical, social and political challenges of artificial intelligence in health*, a report by Future Advocacy for the Wellcome Trust, 2018.

[65] Natalia Norori *et al.*, «Addressing bias in big data and AI for healthcare: a call for open science», *Patterns*, 2(10) 100347, 2021.

avoid causing harm. Furthermore, any compensation should also reflect the degree of damage experienced. However, establishing accountability in an AI liability framework could be difficult in cases in which certain circumstances or legal jurisdictions may present elements preventing individuals from claiming compensation. For example, in the United States, a patient who relies on an AI for medical advice may encounter obstacles in obtaining damages due to the fact that AI systems are not governed by rules of professional liability. Additionally, exclusions or restrictions within product or consumer liability legislation may impede the process of restitution.[66] However, in some areas of healthcare, compensation may be provided without establishing fault or liability, as observed in instances of medical harm due to negative effects from vaccinations. The initial WHO guidance suggested exploring the suitability of «whether no-fault, no-liability compensation funds are an appropriate mechanism for providing payments to individuals who suffer medical injuries due to the use of AI technologies, including how to mobilize resources to pay any claims.»[67]

The World Health Organization has determined that it is necessary for governments to create a comprehensive liability framework covering the entire chain of events from development to deployment of big language models. This structure would allow those harmed by AI technology to claim compensation, despite the complexities involved in pinpointing fault and defining responsibilities among the numerous parties involved in technology's development and use. Furthermore, it is crucial to collect data and insights on AI's use in a methodical way, so that it will be feasible to understand how AI is influencing medical processes, regularly evaluate safety risks based on real-world data, and understand how patients, doctors, and AI systems have adapted to this technology.

It is clear that AI is neither inherently responsible nor irresponsible; rather, its ethical use fully depends on how the datasets, methods, and processes are employed when it is developed, used, or implemented. Policymakers are now in charge of encouraging the ethical use of AI and preventing misuse.

---

[66] Mindy Nunez Duffourc, and Sara Gerke, «The proposed EU directives for AI liability leave worrying gaps likely to impact medical AI», *NPJ Digital Medicine*, 2023; 6(1):77.
[67] World Health Organization, *Ethics and governance of artificial intelligence for health*, 2021.

3.8 Ensure transparency and explainability from AI systems providers and in the relationships between healthcare professionals and patients

As we have explored earlier, the complexity of dynamic machine learning models and obscure mechanisms are central concerns when evaluating a medical device's safety and efficacy, despite the absence of standards mandating human interpretability.[68] Furthermore, the widespread use of generative AI and non-transparent AI instruments in research can lead to biases and errors that may compromise the accuracy of scientific findings. This suggests that while the generated outputs could skew scientific truths, they still seem credible. Even as policymakers and academics seriously debate whether current laws can handle AI, there's a growing list of AI health applications getting the green light. From 2015 to 2020, the US gave the nod to over 222 AI-powered medical tools, with Europe approving 240.[69]

Over time, the notion of transparency has undergone a transformation, increasingly being viewed as a set of measures instead of a singular duty. Currently, this perspective is gaining traction, as already discussed, as the fundamental principle for the advancement and application of artificial intelligence.[70] Examining the principal EU policy documents about AI regulation (AI HLEG Ethics Guidelines,[71] the White Paper on AI,[72] the EP Report on AI Framework,[73] and the Artificial Intelligence Act) one can observe that the components of «transparency» are quite uniform and encompass aspects like explainability, interpretability, means of communication, auditability, traceability, providing information, record-keeping, and documentation. Moreover, the AI Act delineates transparency obligations more specifically, indicating comprehensive requirements for providing information.[74] This suggests that transparency is no longer a singular notion

---

[68] Pantelis Linardatos *et al.*, «Explainable AI: A Review of Machine Learning Interpretability Methods», *Entropy*, 23(1), 2020.

[69] Urs J. Muehlematter, Paola Daniore, and Kerstin Vokinger, «Approval of artificial intelligence and machine learning-based medical devices in the USA and Europe (2015–20): a comparative analysis», *Lancet Digit Health*, 3(3), pp. e195–e203, 2021.

[70] Anastasiya Kiseleva, *et al.*, «Transparency of AI in healthcare as a multilayered system of accountabilities: between legal requirements and technical limitations», *Frontiers in Artificial Intelligence*, 5(879603), 2022.

[71] European Commission, *Ethics Guidelines for Trustworthy AI, High-Level Expert Group on AI*, 2019.

[72] European Commission, *White Paper on Artificial Intelligence: a European approach to excellence and trust*, 2020.

[73] European Parliament, *Report with Recommendations to the Commission on a Framework of Ethical Aspects of Artificial Intelligence, Robotics and Related Technologies—Report—A9-0186/2020*, 2020.

[74] Anastasiya Kiseleva *et al.*, *op. cit.*

or mandate but comprises various initiatives that must be adopted to achieve it, such as managing information or overseeing data.

Although the various strategies and provisions adopted in the AI Act, and extensively discussed in Chapter 4 may appear disparate and supplementary to the need for transparency, they are in fact a collection of protocols that make transparency achievable. Without access to information and data, the assurance of transparency is not possible, and understanding how an AI system functions is vital for maintaining its safety and integrity.[75] This strategy aligns with the framework established by the GDPR: While the Regulation aims to implement the principle of transparency, it also includes several safeguards to ensure adherence, such as provision of explanation of the decision based solely on automated processing and producing substantial effects on data subjects (GDPR, recital 71); informed consent requirement (GDPR, art. 7); requirements of information provision and communication imposed on data controllers (GDPR, art. 12-15); obligation to provide access to personal data upon request of data subject (GDPR, art. 15); records of processing activities (GDPR, art. 30).

Tied to the principle of transparency is the medical informed consent requirement,[76] which guarantees patients' knowledge and autonomous decision-making, upholding their right to dignity. Are healthcare professionals required to inform patients when AI/ML systems are being utilized? How much disclosure is necessary? Does their responsibility shift if they decide not to follow the AI's recommendation? In cases where an AI system is a black box, is there an obligation to reveal that its reasoning cannot be entirely understood or clarified? Is it the duty of healthcare providers to enlighten patients about how AI works? Other challenges posed by AI include the use of health applications and chatbots that, due to their reliance on user agreements rather than direct interaction, complicate the conventional process of obtaining informed consent.[77]

It is crucial for healthcare professionals and patients to have the necessary tools for making informed decisions. In their interactions with patients, doctors must provide clear and thorough explanations that enable patients to understand and participate in their own care. However, the challenge of achieving informed consent extends into the digital age with AI integration,

---

[75] Sara Gerke *et al*, «Ethical and legal challenges of artificial intelligence-driven healthcare», *Artificial Intelligence in Healthcare*, 2020, pp. 295-336.

[76] I. Glenn Cohen, «Informed consent and medical artificial intelligence: what to tell the patient?», *The Georgetown Law Journal*, 108(1425), 2020, pp. 1425-1469.

[77] Craig M. Klugman *et al.*, «The ethics of smart pills and self-acting devices: autonomy, truth-telling, and trust at the dawn of digital medicine», *The American Journal of Bioethics*, AJOB, 18(9), 2018, pp. 38-47.

where it's not just about a healthcare provider's medical knowledge but also their grasp on AI technologies that support decision-making. Physicians require comprehensive information regarding when and how to deploy AI mechanisms effectively and how to verify the outcomes prompted by these intelligent systems.

3.9 Can artificial intelligence be an inventor?

Historically, only humans could be recognized as inventors. In the United States, this concept was reiterated in 2022 in Thaler v. Vidal, No. 21-2347 (Fed. Cir.), when the Federal Circuit relied on the Patent Act's definition of an «inventor, » which is identified as «the individual or, if a joint invention, the individuals collectively who invented or discovered the subject matter of the invention,» ruling out the possibility of granting inventorship to AI, which is clearly not a human being.[78] In Europe, the Legal Board of Appeal at the European Patent Office (EPO) came to the same conclusion, stating that AI cannot be designated as an inventor on patent filings.[79] Furthermore, in December 2023, the UK Supreme Court unanimously determined that also under the UK. Patent law only natural persons can be named as inventors, hence excluding AI entities.[80] Presently, South Africa stands as the only nation that has recognized an AI as an inventor,[81] a decision that has prompted significant global controversy and that may be explained by South Africa's eagerness to use AI to its full potential to try to solve some of the country's socioeconomic issues.

As it stands, it's evident that a human inventor may collaborate with an AI during the invention process, yet the AI in question cannot be credited as an inventor on any resulting patent. In this case, complexities may emerge regarding shared invention contributions for AI-aided inventions. For these situations, drawing from the insights of the Thaler case, we can deduce that in order to secure a patent for example under the United States Patent Act, a human working alongside AI must still provide a contribution of legal significance sufficient to qualify as a coinventor according to the statute. This means that a human inventor utilizing a generative AI will have to make

---

[78] US Patent Application Nos. 16/524,350 and 16/524,532 (filed July 29, 2019).

[79] Patent Application Nos. EP 18 275 163, concerning a *Food Container* and EP 18 275 174, relating to *Devices and Methods for Attracting Enhanced Attention* (filed October 17, 2018, and November 7, 2018).

[80] Patent Application Nos. GB1816909.4 and GB1818161.0 (filed October 17, 2018, and November 7, 2018).

[81] Patent Application No. PCT/IB2019/057809 (filed September 17, 2019).

significant contributions either through inputting training data into the AI system or by refining the output.

This situation introduces novel considerations, essentially prompting legislators to delineate what degree and nature of human involvement is necessary for inventor recognition, as well as the evidence required to substantiate such human input. Any choice that will be made by policymakers and legislators regarding this aspect will affect profoundly patent law and the global panorama of inventorship, delineating new scenarios in innovation and research.

3.10 Promoting AI literacy and education between healthcare professionals, patients, and institutions

Today's policymakers are tasked with the responsibility of strategically investing in formal education or leveraging existing learning structures to improve AI literacy among citizens, which is essential for ensuring ethical oversight, diminishing biases, and enhancing public trust in the fairness of AI systems.

In order to promote the responsible use of AI in educational settings, the Executive Order issued by US President Joe Biden in October 2023 on «Safe, Secure, and Trustworthy Development and Use of Artificial Intelligence» directed the Education Department to create an AI toolkit. This resource aimed to aid schools in implementing best practices in AI usage, emphasizing the importance of data privacy, reducing bias, and safeguarding young people. Nonetheless, in order to be really effective, all these initiatives would still require further implementation through additional measures, and for this reason, the US Congress has been particularly active in proposing legislation to address these issues. For example, Rep. Lisa Blunt Rochester (D-DE) and Rep. Larry Bucshon (R-IN) proposed Bill 6791 to the US House of Representatives in December 2023, titled the Artificial Intelligence Literacy Act. The bill considers AI literacy as a fundamental aspect of digital proficiency, and as a necessity for national competitiveness. This legislation quickly earned the backing of educators' associations, higher education institutions, tech companies, industry organizations, and nonprofits engaged in educational or occupational training.

AI education becomes particularly important in healthcare, where multiple individuals interact with AI models and with each other, with the need to understand AI processes and outputs in order to make a decision regarding the health and life of a patient or of themselves. Transparency of AI usage in clinical settings is essential to involve all stakeholders, paving the way for its ethical and responsible employment and to empower healthcare professionals with the possibility of making informed decision-making capabilities, while allowing patients to engage actively in their own treatment decisions.

Many policies can be put in place to enhance AI literacy, as, for instance, embedding AI education into the curriculum of medical schools. In fact, current research[82] indicates a lack of thorough understanding of AI technologies among most students and faculties in these institutions. Yet, there exists a noticeable eagerness to learn more about AI across various domains. Interestingly, despite the heavy focus on knowledge acquisition due to the wide-ranging subjects covered in biomedical and clinical studies (which also leaves limited resources for additional instruction), the consensus remains that AI education should be an integral part of medical training, suggesting that it would be beneficial to start teaching it in medical schools.

However, AI literacy shouldn't be confined to the healthcare professionals alone; it is, in fact, equally imperative to implement strategies to educate patients. This is for mainly two reasons: Primarily, patients need clarity on how AI impacts their diagnosis and treatment plans in order to make informed decisions about their healthcare; additionally, educating them provides the necessary skills to understand and evaluate the importance of data privacy, consent, and responsible AI usage. Moreover, AI literacy is intrinsically linked to health literacy, as AI can provide patients with the resources to better understand and manage their health conditions, also contributing to addressing health disparities.[83] In this context, AI can and should serve as a tool for translating complex medical concepts in a more accessible language, guaranteeing that all patients, including those coming from underserved communities, have access to comprehensive and comprehensible health information.

3.11 Measuring the real environmental impact of artificial intelligence

In 2022, worldwide electricity usage by data centers was estimated at around 460 terawatt-hours and is projected to potentially double by 2026, surpassing

---

[82] Elena A. Wood *et al.*, «Are we ready to integrate artificial intelligence literacy into medical school curriculum: students and faculty survey», *Journal of Medical Education and Curricular Development*, 8(23821205211024078), 2021.

[83] Center for Advancing Safety of Machine Intelligence—a collaboration between Northwestern University and UL Research Institutes, *How Researchers are Using Artificial Intelligence to Help Low Literacy Patients Understand Their Health*, 2023.

1,000 TWh, which is roughly equivalent to the entire power consumption of Japan. Also, a single ChatGPT query requires 2.9 watt-hours of electricity, compared with 0.3 watt-hours for a Google search.[84] However, the environmental footprint of artificial intelligence is difficult to quantify, particularly due to the lack of formal regulations and standardized methods for calculating the emissions attributable to AI.[85] Additionally, as AI models grow in complexity, their energy demands for training and operation increase.

Preliminary findings from studies conducted by researchers at Hugging Face and Carnegie Mellon University indicate that the energy required to generate a single image with a robust AI model is comparable to the energy used to charge a smartphone completely.[86] Furthermore, training a standard AI language-processing system can result in the production of approximately 1,400 pounds of carbon emissions, which equates to the emissions generated by one roundtrip flight between New York and San Francisco for an individual. The same research showed that the complete set of experiments required to construct and train an AI language model from the beginning can produce as much as 78,000 pounds of CO_2, which is double the amount the average American breathes out in their whole life.[87] We should also consider the significant water usage associated with artificial intelligence: Beyond the water used for electricity production, in fact, copious additional amounts of clean water are essential for cooling the heatsinks of AI servers to forestall overheating. This cooling is typically achieved using cooling towers, open-air systems, and large volumes of water.

When examining the entire AI supply chain, other stages may also demand substantial water use (for instance, manufacturing a single microchip requires roughly 2,200 gallons of Ultra-Pure Water (UPW)). Furthermore, training a sophisticated language model like GPT-3 involves the use of millions of liters of water, while just performing 10–50 queries with GPT-3 inference can result in 500 milliliters of water usage, which varies based on the temporal and geographical location of the hosting server. Subsequent iterations, such as GPT-4, are said to be even larger and thus potentially more water-intensive than GPT-3.[88]

---

[84] International Energy Agency, *Electricity 2024*, 2024.

[85] David Berreby, *As Use of A.I. Soars, So Does the Energy and Water It Requires*, Yale Environment 360, February 6, 2024.

[86] Melissa Heikkila, «Making an image with generative AI uses as much energy as charging your phone», *MIT Technology Review*, 2023.

[87] Edmund L. Andrews, *AI's Carbon Footprint Problem*, Stanford Human Centered Artificial Intelligence, July 2, 2020.

[88] Shaolei Ren, *How much water does AI consume? The public deserves to know*, OECD.AI Policy Observatory, November 30, 2023.

This situation raises several concerns: First, issues related to resource allocation, as AI's growing water use competes with necessities like drinking water and agriculture. Then, as water's temperature is usually higher when it is released back into the environment, there may be repercussions on regional ecosystems.

Policymakers in the world are starting to look for solutions to make AI more environmentally responsible. For instance, in early 2024, US Representatives Eshoo and Beyer, along with Senators Markey and Heinrich, brought forth the Artificial Intelligence Environmental Impacts Act, which aimed to increase awareness of this issue and start exploring potential solutions. The proposed legislation would also ensure that the Environmental Protection Agency performs an extensive review of AI's environmental effects within two years. Additionally, the National Institute of Standards and Technology would be required to gather a group of interested parties to create standards for assessing the environmental consequences of AI. Sen. Edward Markey, again, who was also at the time chairman of the Environment and Public Works Committee panel on clean air, climate, and nuclear safety, repeatedly expressed his intention to halt any forthcoming bills related to artificial intelligence that do not comprehensively consider the technology's environmental impact.

On one hand, the public should be made aware of the environmental impact of AI, while simultaneously, policymakers in all countries and international organizations should build a framework for gathering in-depth information about such impacts. Policymakers are in fact now presented with a significant challenge: Without concrete metrics to gauge the influence of AI, formulating effective strategies, and addressing issues is nearly unattainable.[89]

On the other hand, one could hope that AI's carbon footprint may diminish over time as technologies advance in efficiency and new methods are discovered, but given that its use would be more spread, that is a hard bet. However, it can be highlighted that AI has the potential to be instrumental in combating climate change and reducing pollution, maybe being the solution to some of the problems enhanced by itself. This could happen, for example, through the use of AI to develop energy-efficient architecture, to enhance the analysis of satellite data for Earth monitoring, identifying, and planning against deforestation, or to improve the implementation of renewable energy sources.

---

[89] Jesse Dodge, *et al.*, «Measuring the Carbon Intensity of AI in Cloud Instances», in *Proceedings of the 2022 ACM Conference on Fairness, Accountability, and Transparency* (FAccT '22). Association for Computing Machinery, New York, NY, USA, June 20, 2022, pp. 1877-1894.

PART III
REGULATING ARTIFICIAL INTELLIGENCE WORLDWIDE: DIFFERENT APPROACHES FOR ADDRESSING DIFFERENT NEEDS AND PRIORITIES

4 Safeguarding Individuals and Society as the Main Priority: The European Right-based Approach

4.1 Introduction to the European regulatory approach to AI

Regulating the development and utilization of AI is essential for adequately safeguarding both the people and stakeholders involved, and in particular their fundamental rights that may be jeopardized by its application. In doing so, as mentioned in the previous pages, the final aim of the legislation includes ensuring a sufficient level of trust in the technology, an essential pre-requisite for the long-term successful use of AI in any field, but especially in healthcare.

Creating an ecosystem of trust is precisely one of the fundamental goals of the European approach to regulating AI, alongside establishing an «ecosystem of excellence» throughout the entire value chain and life cycle of AI. The European Commission's strategy to achieve both goals involves adopting regulations that govern the design, development, and use of AI. Simultaneously, it aims to protect European values and provide legal certainty by identifying the necessary norms and measures to ensure compliance with them.[1] In doing so, the European legislators aimed to maintain a human-centric approach, with special consideration to the principles delineated in the Communication on Building Trust in Human-Centric Initiatives AI of the European Commission.[2]

The regulatory pathway on the matter originated with the Resolution of the European Parliament directed toward the Commission, proposing the development of regulatory solutions in the domain of civil law pertaining to robotics in 2017. This discussion subsequently reached a significant point with the publication of the White Paper on Artificial Intelligence in 2020, which assessed and outlined various policy options in this area.[3]

---

[1] European Commission, «White paper on artificial intelligence—a European approach to excellence and trust», 2019.

[2] Communication from the Commission to the European Parliament, the Council, the European Economic and Social Committee and the Committee of the Regions Building Trust in Human-Centric Artificial Intelligence, COM/2019/168 final, 2019.

[3] For an extensive analysis of the overall process that led to the approval of the

Upon consideration of the necessity for a precise and legally binding document to achieve the aforementioned objectives, it was evident that ethical standards and self-regulatory policies alone were inadequate to this end.[4] Therefore, after extensive consultations with the stakeholders in every stage of the legislative journey,[5] the Proposal for a Regulation on AI was presented by the European Commission in April 2021, and the final text of Regulation 2024/1689 was published in the Official Journal of the European Union on July 12, 2024,[6] marking a significant advancement in the international race for the regulation of AI.[7]

The Proposal explicitly aimed at implementing the ecosystem of trust envisioned in the White Paper[8] and aspired to guarantee the European Union's role as «a global leader in the development of secure, trustworthy, and ethical Artificial Intelligence.»[9] By adopting the legislative form of a regulation, the AI Act introduces binding harmonized standards and norms to be complied with for the marketing and use of AI systems in the Union market.

Other key characteristics of the regulatory approach adopted by the European Union may be identified. In particular, the AI Act is a *horizontal product regulation*[10] and is part of the New Legislative Framework first established in 2008. This choice of regulating AI merely as a product, like any other product in the European market, from the simplest to the most advanced, has been at times criticized for oversimplifying the complex reality of AI.[11]

---

final version of the AI Act, see Nikos Th. Nikolinakos, *EU policy and legal framework for artificial intelligence, robotics and related technologies—The AI Act*, Springer, 2024.

[4] Dimitar Lilkov, «Regulating artificial intelligence in the EU: a risky game», *European View*, 20(2), 2021, pp. 166-174.

[5] For a precise analysis and description of the regulatory reform process and legislative process that led to the development of the Proposal and finally to the adoption of the AI Act, Ronit Justo-Hanani, «The politics of artificial intelligence regulation and governance reform in the European Union», *Policy Sciences*, 55, 2022, pp. 137-159.

[6] Regulation (EU) 2024/1689 of the European Parliament and of the Council of June 13, 2024 laying down harmonized rules on artificial intelligence and amending Regulations (EC) No 300/2008, (EU) No 167/2013, (EU) No 168/2013, (EU) 2018/858, (EU) 2018/1139 and (EU) 2019/2144 and Directives 2014/90/EU, (EU) 2016/797 and (EU) 2020/1828 (Artificial Intelligence Act).

[7] Rostam J. Neuwirth, «Prohibited artificial intelligence practices in the proposed EU Artificial Intelligence Act», *Computer Law & Security Review*, 48, 2023, pp. 1-14.

[8] In these terms, the Explanatory Memorandum to the Artificial Intelligence Act by the European Commission, COM(2021) 206 final 2021/0106(COD).

[9] European Council, Special meeting of the European Council (1 and 2 October 2020)—Conclusions, EUCO 13/20 (2 October 2020) paragraph 13.

[10] Explanatory Memorandum to Artificial Intelligence Act, European Commission, paragraph 1.3.

[11] Edwards Lilian, «Expert opinion—regulating AI in Europe: four problems and four solutions», 2022, available at https://www.adalovelaceinstitute.org/report/regulating-ai-in-europe/.

The newly enacted regulation builds upon a wide variety of existing laws, which may be identified by the interpreter considering, on the one hand, the variety of possible applications of an AI system and thus its potential additional qualifications as a product otherwise regulated under Union law, and on the other considering that for its development and use various techniques and activities come into consideration, all of which are in turn addressed by different norms. As a result, putting together «the panoply of technological regulations in EU law, admittedly, is no easy task»[12] and will take the form of a legislative square (or pentagon).[13]

Concerning the healthcare sector specifically, the regulations necessitating consideration and, where applicable, compliance can be divided into three principal categories: (I) the standards related to the manufacturing and marketing of products employed in the provision of healthcare, (II) the regulatory framework overseeing the processing of personal data along with the protection of privacy rights and data security, and (III) the guidelines pertaining to liability in instances involving the application of artificial intelligence.

Regarding the initial set of standards (I), it is imperative to reference to Regulation 2017/745 (Medical Device Regulation—MDR) and Regulation 2017/746 (In-Vitro Diagnostic Regulation—IVDR), which establish norms applicable to *medical devices* and *in-vitro medical devices*, respectively. As elaborated upon in the subsequent pages, numerous artificial intelligence systems designed for medical or clinical applications will be subject to both the AI Act and either the MDR or the IVDR. Indeed, AI systems will fall under the first category when designed and intended by the provider to «be used, alone or in combination, for human beings for one or more specific medical purposes»[14] listed by the regulation itself. Differently, they should be qualified as in-vitro devices whenever «intended by the manufacturer to be used in vitro for the examination of specimens, including blood and tissue donations, derived from the human body, solely or principally for the purpose of providing information» on a set of identified topics.[15] In

---

[12] Ugo Pagallo, «Why the AI Act won't trigger a Brussels effect, (December 16, 2023)», *AI Approaches to the Complexity of Legal Systems*, Springer, 2024, available at SSRN: https://ssrn.com/abstract=4696148; Ugo Pagallo, Jacopo Ciani Sciolla, and Massimo Durante, «The environmental challenges of AI in EU law: lessons learned from the Artificial Intelligence Act (AIA) with its drawbacks», *Transforming Government: People, Process and Policy*, 16(3), 2022, pp. 359-376.

[13] Luciano Floridi, «Translating principles into practices of digital ethics: five risks of being unethical», *Philosophy & Technology*, 32(2), 2019, pp. 185-193.

[14] Article 2 paragraph 1 MDR.

[15] Article 2 paragraph 2 IVDR.

the following pages, we will discuss AI as a medical device, and therefore the provisions established by the MDR, considering that the latter will be the regulation applicable in the vast majority of cases where AI is applied in healthcare.

The second component (II) is predominantly constituted by Regulation (EU) 2016/679, known as the General Data Protection Regulation (GDPR), in conjunction with pertinent soft law and interpretative tools. This framework is further supplemented by additional legislative measures outlined in the European Strategy for Data, specifically including Regulation (EU) 2022/868, referred to as the Data Governance Act, Regulation (EU) 2023/2854, known as the Data Act, and Regulation 2025/327 on the European Health Data Space (EHDS).

Finally, (III) when considering liability in the context of AI-assisted care, one must account for the newly established Directive 2024/2853 on liability for defective products (Product Liability Directive), which replaces the former Council Directive 85/374/EEC. In addition, from 2022 until February 2025 attention was directed toward the Proposal for a Directive that aims to adapt non-contractual civil liability rules to artificial intelligence (AI liability directive), which however was abandoned in early 2025. Notwithstanding this, it is worth analyzing the text of such proposal, in order to describe the approach taken by the European Union on the matter, and for this reason, we included a specific section on the AI liability directive in this volume.

All the mentioned regulations and norms shall be taken into consideration and, in most cases, simultaneously applied when manufacturing and using an AI in the EU, in the sense and with the difficulties that will be analyzed in the following pages, which aim to provide an overview of the complex task of navigating the European regulatory waters and being compliant with the normative framework applicable to AI.

4.2 The risk-based approach of the AI Act

In regulating AI in the Union market, the European co-legislators adopted a risk-based approach, as explicitly stated by the European Commission already in the Explanatory Memorandum to the Proposal.[16] Such an approach, which has been addressed as «the right approach to AI regulation»,[17] is rooted in the belief shared with the experts in the field and almost any soft law instruments on the matter (and already underlined in the previous pages) that the design, marketing, and use of AI systems may hinder fundamental human rights and have unprecedented consequences for Union values and principles.[18] At the same time, however, these systems have the potential to drastically improve people's quality of life[19] because of the tremendous innovations that their use in healthcare may bring.

As a consequence, the aim of the AI Act, clearly stated at the beginning of both the Explanatory Memorandum to the Proposal and of the Regulation itself, was precisely «to improve the functioning of the internal market (...) in accordance with Union values, to promote the uptake of human-centric and trustworthy artificial intelligence (AI) while ensuring a high level of protection of health, safety, fundamental rights as enshrined in the Charter (...), including democracy, the rule of law and environmental protection, to protect against the harmful effects of AI systems in the Union, and to support innovation».[20] In doing so, the regulation thus pursues a traditional constitutional goal: that of striking a fair balance between (possibly) competing fundamental rights and interests worthy of protection according to Union law. Once again, in the context of AI, this would mean balancing the interests of companies and investors involved in the research on and development of the technology on the one hand and those of patients and society possibly affected by its use on the other.

Adopting a risk-based approach to regulating AI is opposed to using merely an ex-post liability framework, i.e., norms that, instead of regulating how an AI system should be developed, are designed to address directly the

---

[16] *Ibid.*, paragraph 1.1, «The proposal sets harmonised rules for the development, placement on the market and use of AI systems in the Union following a proportionate risk-based approach».

[17] Martin Ebers, «Truly risk-based regulation of artificial intelligence—how to implement the EU's AI Act», *European Journal of Risk Regulation*, 2024, pp. 1-20.

[18] Explanatory Memorandum to Artificial Intelligence Act, European Commission, paragraph 1.1.

[19] White Paper on Artificial Intelligence—A European approach to excellence and trust, European Commission, February 19, 2020.

[20] Recital 1, AI Act.

legal consequences of a wrongdoing caused by (the use of) an AI. The one of the European Commission in the AI Act is thus a precise «value-laden choice»[21] that has the merit of anticipating and attempting at avoiding harms from the outset by establishing categories of AI systems based on the risks posed by their use and calibrating requirements accordingly. One reason for the basis of this approach is that harms in the context of AI may be nearly impossible to clearly envision, are often «latent in nature» and may be particularly difficult and burdensome to prove in court after they occurred, due in particular to the issues such as the opaqueness of the system, black box AI, etc., and in terms of costs of the eventual litigation.[22] At the same time, however, anticipating and regulating possible failures of the systems at the moment of enacting the norms is a complex normative exercise, given the high technical nature of the matter and the (at times) unpredictability of the functioning and results of the technology itself.

This approach is far from being new.[23] Indeed, it is the one usually chosen for other product regulations already enacted and fully applicable in Europe, especially within the New Legislative Framework, and has recently been adopted also in the context of the protection of personal data and the right to privacy by the GDPR,[24] as well as more generally in that of the EU digital policies.[25]

However, while in principle the aim of these regulations is always that of mitigating the risks posed to constitutional values by the activity or the product regulated by the enacted norms, the concrete way to achieve this goal may differ. Indeed, at least two main approaches may be theorized: the *top-down* approach and the *bottom-up* approach.[26] While both architectures pursue the goal of providing a system that guarantees the maximum level of protection of fundamental values and the minimum level of harm, they radically differ in the degree of specificity of the norms and, in part consequent-

---

[21] Margot E. Kaminski, «Regulating the risks of AI», *Boston University Law Review*, 103(1347), 2023, p. 1351.

[22] *Ibid.*, p. 1366.

[23] Milda Macenaite, «The 'Riskification' of European Data Protection Law through a two-fold Shift», *European Journal of Risk Regulation*, 8(3), 2017, pp. 506–540.

[24] Regulation (EU) 2016/679 of the European Parliament and of the Council of April 27, 2016, on the protection of natural persons with regard to the processing of personal data and on the free movement of such data, and repealing Directive 95/46/EC (General Data Protection Regulation).

[25] Also, for instance, the Digital Service Act. Giovanni De Gregorio, and Pietro Dunn, «The European risk-based approaches: connecting constitutional dots in the digital age», *Common Market Law Review*, 59(2), 2022, pp. 473-500.

[26] *Ibid.*

ly, in that of the freedom of the natural (or legal) persons to which they apply *ratione personae* to choose the measures most suitable to ensure compliance, as well as the type and amount of duties entrusted to them.

The General Data Protection Regulation (GDPR) recently embraced the bottom-up approach to safeguarding the fundamental right to privacy and data protection for natural persons. Within this framework, the law delineates the interest to be protected and the principles that must be adhered to while delegating the responsibility of selecting the most effective concrete measures for achieving the regulatory objectives to the entities governed by the regulation. In the context of the GDPR, these entities primarily include the data controller, followed by the data processor.

Article 5 of the GDPR clearly encodes this structure in the principle of accountability, according to which it is a duty of the data controller to evaluate the concrete risks of the data processing to be conducted, to have in place appropriate technical and organizational measures to mitigate them, and to be able to demonstrate the effectiveness of the implemented system upon request. To reinforce the described system, as well as facilitate and ensure compliance with the norms, various instruments are provided for by the GDPR, such as the data protection impact assessment (DPIA) in Article 35, i.e., «an assessment of the impact of the envisaged processing operations on the protection of personal data,» for specific types of processing that are «likely to result in a high risk to the rights and freedoms of natural persons,» such as when new technologies are used, and taking into account the concrete nature, scope, context, and purpose of the processing.[27]

On the other hand, the top-down approach is the one usually adopted by European product regulations. In these scenarios, it is the law itself that identifies the risks posed by a given type of product (or technology) to fundamental rights and values and consequently categorizes the products based on this evaluation. Norms and obligations provided for by the regulation and to be complied with are then weighted by the legislator accordingly. Therefore, the economic operators at various stages and in different degrees involved in the design, production, and marketing of products that fall under the scope of application of this kind of regulation are not responsible for preliminarily and autonomously identifying the risks related to their product or technology, and thus for adopting measures (technical or otherwise) to mitigate such risks, but should identify under which category their product falls and comply with the corresponding (already defined) obligations. In particular, this is especially true for the economic operator that manufactures the

---

[27] Article 35, GDPR.

AI system, which is specifically addressed and defined in the AI Act as the *provider*. Indeed, it is the provider that primarily will have to identify the concrete requirements and norms to be complied with for developing the AI system (or having it developed under their responsibility) and placing it on the Union market «under its own name or trademark, whether for payment or free of charge.»[28]

The described top-down approach is the one adopted, for instance, by Regulation 2016/425 for personal protective equipment (PPE) or the MDR for medical devices, to name but a few examples. For both types of products (PPE and medical devices), the European co-legislators identify different classes based on their risk. For instance, article 18 of Regulation 2016/425 explicitly states that «[t]he PPE shall be classified according to the risk categories set out in Annex I,» which in turn provides for a list of risks against which the PPE is intended to protect the users, divided into minimal risks (PPE category I), risks that may cause very serious consequences such as death or irreversible damage to health (PPE category III) and other risks (PPE category II). Similarly, article 51 of the MDR establishes that «[d]evices shall be divided into classes I, IIa, IIb, and III, taking into account the intended purpose of the devices and their inherent risks.»

Coherently with the choice of regulating AI as a product, the AI Act replicates the same top-down scheme adopted by Regulation 2016/425 and the MDR. In particular, the Act aims to provide norms for both AI systems and general-purpose AI systems, which are classified according to the concrete possible risks that may arise from their design, marketing, and use on the Union market and territory. To this end, Article 3, paragraph 2 of the AI Act defines *risk* as «the combination of the probability of an occurrence of harm and the severity of that harm.»

As for AI systems, the categories identified are (1) AI systems that pose an *unacceptable risk* and that as a consequence are prohibited on the European market; (2) *high-risk* AI systems that may be marketed and used provided that specific technical requirements are complied with; (3) *limited-risk* AI systems, for which the regulation only enacts some obligations on the provider and the deployers; and (4) AI systems that pose *minimal* or *no risk*.[29] In addition, during the negotiation of the initial Proposal for the AI Act, the European co-legislator decided to include a new category of AI systems,

---

[28] Article 3 paragraph 3, AI Act.
[29] Among others, Martin Ebers, «Truly risk-based regulation of artificial intelligence—how to implement the EU's AI Act», *op. cit.*, p. 1; Claudio Novelli *et al.*, «AI risk assessment: a scenario-based, proportional methodology for the AI Act», *Digital Society*, 3(1), 2024, pp. 1-29.

namely the «general-purpose AI» models and systems, for which a similar (but not equal) approach was adopted. Indeed, general-purpose AI models are divided into (I) those that present *systemic risk* (general-purpose AI systems *with systemic risk*) and (II) those that do not (general-purpose AI systems). General-purpose AI systems with systemic risks are those that present «high-impact capabilities» and thus «capabilities that match or exceed the capabilities recorded in the most advanced general-purpose AI models,» evaluated «on the basis of appropriate technical tools and methodologies, or significant impact on the internal market due to its reach.»[30]

The risk pyramid[31] or categorization so identified, both for «ordinary» AI systems and general purpose, ensures that norms and obligations are applicable only in so far as the technology under consideration poses an actual risk to protected rights and values, with the degree of severity of the legislation being directly proportional to the level of threat or risk, according to the proportionality principle.[32] In this regard, the AI Act is coherent with the other regulations generally enacted in the field of technology and science in aiming at finding the appropriate balance by establishing norms to safeguard fundamental values and rights on the one hand, while not excessively constraining possible developments of the technology itself and the European economy on the other.

4.2.1 The product categories identified by the AI Act—(1) prohibited AI systems

The AI Act provides for a very precise, descriptive, and (in the intention of the co-legislators) comprehensive list of AI systems that are considered as posing an unacceptable risk to the fundamental rights and values protected by Union law,[33] and therefore that cannot be manufactured, placed on the market or even used in the European Union. In this regard, the AI Act widens considerably the scope of the prohibition in Article 5 by limiting not only the marketing of various types of AI, but also their use. As a consequence,

---

[30] Recital 111, AI Act.

[31] Dimitar Lilkov, *op. cit.*

[32] Recital 26 «In order to introduce a *proportionate and effective* set of binding rules for AI systems, a clearly defined *risk-based approach* should be followed. That approach should *tailor the type and content of such rules to the intensity and scope of the risks* that AI systems can generate. It is therefore necessary to prohibit certain unacceptable AI practices, to lay down requirements for high-risk AI systems and obligations for the relevant operators, and to lay down transparency obligations for certain AI systems.» (emphasis added). Recital 26 thus reflects the principle of proportionality enshrined in Article 5 paragraph 4 TFEU.

[33] Explanatory Memorandum to Artificial Intelligence Act, European Commission, p. 12 (pt. 5.2.2).

both the provider and the deployer may be sanctioned for violating the mentioned article, for having placed a prohibited AI system on the market or having used it for its intended purpose or otherwise, respectively. Differently from most of the provisions of the AI Act, this prohibition is already applicable from February 2, 2025.

While originally Article 5 of the Proposal was limited to *subliminal* AI practices, AI practices that *exploit vulnerabilities, social scoring* AI systems, and *biometric* AI systems,[34] the final version doubled the prohibited kinds of AI, thus attempting to increment also the instances of protection. Notwithstanding the fact that most of the AI systems listed in Article 5 are (and will) not generally (be) used in the field of healthcare, a brief description of them may be useful to understand the general approach of the European co-legislator toward regulating AI.

The first group of prohibited AI comprises *subliminal* AI systems that «[deploy] subliminal techniques beyond a person's consciousness or purposefully manipulative or deceptive techniques» and are prohibited as long as they are used with the «objective, or the effect of materially distorting the behaviour of a person or a group of persons by appreciably impairing their ability to make an informed decision, thereby causing them to take a decision that they would not have otherwise taken in a manner that causes or is reasonably likely to cause that person, another person or group of persons significant harm.»[35] Moreover, and similarly, the second category includes AI systems that *exploit the vulnerabilities of natural persons* (or specific group of persons) «due to their age, disability or a specific social or economic situation.» In this case as well, the AI system shall aim at «materially distorting the behaviour of [that person] in a manner that causes or is reasonably likely to cause that person or another person significant harm.»[36]

For both the described categories of AI systems, the AI Act seems to have acknowledged the need to ensure compliance with the ethical principle of the respect of human autonomy by prohibiting, in legal terms, all those systems that may hinder and endanger a full expression of individual self-determination, and that of the prevention of harm.

A third group encompasses all artificial intelligence systems utilized to partially classify natural persons or infer or predict their behavior based on personal or sensitive information, which may result in potential discriminatory effects. This category includes AI systems employed for the evaluation

---

[34] For an in-depth analysis of the firstly proposed version of article 5, see Rostam J. Neuirth, *op. cit.*, p.1.

[35] Article 5, letter a, AI Act.

[36] Article 5, letter b, AI Act.

or classification of natural persons or groups of individuals over an extended period, grounded on their social behavior or known, inferred, or predicted personal or personality characteristics, commonly referred to as *social scoring* systems. These systems are prohibited only to the extent that the social score derived results in detrimental or unfavorable treatment of certain individuals in social contexts that are unrelated to those in which the data processed by the AI system was originally generated or collected, or, as an alternative or in addition, if such treatment is unjustified or disproportionate in relation to their social behavior or its gravity.

Moreover, also prohibited are AI systems implemented for facial recognition purposes in order to create databases[37] or for inferring emotions in the workplace and in educational institutions, with the exception in the last scenario of when these systems are used for medical or safety reasons. Finally, Article 5 lists AI systems used to analyze biometric data, either in order to categorize natural persons or in real-time in publicly accessible spaces for law enforcement purposes, with the exception in both cases of this data being processed for law enforcement purposes under specific conditions. Differently from the other instances of prohibited AI systems, the AI Act dedicates various definitions to concepts related to the processing of biometric data and to biometric identification systems, aiming at clearly identifying the perimeter of prohibited and permitted activities of this kind, such as real-time remote biometric identification as opposed to post-remote biometric identification systems, which are lawful under certain circumstances. Moreover, for this type of prohibited AI, and with the aim of striking a fair balance, once again, between possibly contrasting interests, the AI Act provides for various exceptions to this ban, such as in cases where real-time remote biometric identification systems are used to prevent imminent threats to life or to identify criminal perpetrators.[38]

All the listed AI systems in Article 5 would contrast with or violate fundamental rights protected by the Charter of Fundamental Rights, in particular those to non-discrimination, data protection, privacy, and the rights of the child, as well as being non-compliant with the ethical principles discussed in the previous pages, thus also potentially undermining the trustworthiness of AI as a whole. As a consequence, and because of the seriousness of the threats possibly posed by these systems, the ban set forth by the AI Act not only applies to the providers of the listed AI, who thus cannot place them on the Union market, but also extends to the deployers who are instructed

---

[37] Article 5, letter e, AI Act
[38] Article 5, paragraph 1, letter h, points I–III, AI Act.

not to use these systems, thus adding a new layer of protection of the natural persons that may possibly be affected by their use.

Recently, the European Commission provided for extensive further guidance on the interpretation of the prohibited practices of AI according to the mentioned Article 5 AI Act. The document is particularly useful not only for interpreting the norms but also because it provides concrete examples of prohibited AI practices that will fall under the prohibition of marketing and use. For instance, it is suggested that AI systems implemented for evaluating persons with the aim of determining if they are entitled to receive essential public assistance benefits and services qualify as prohibited practices if they involve unacceptable social scoring (and provided that the other conditions of Article 5(1)(c) are fulfilled).[39]

4.2.2 The product categories identified by the AI Act—(2) high-risk AI systems

The vast majority of the norms of the AI Act are dedicated to regulating AI systems that are qualified as *high-risk,* meaning that their use on the Union market is recognized as possibly endangering protected fundamental rights and values if not carefully regulated at the normative level and consequently designed, marketed, and used in compliance with the enacted norms at the concrete one. This category is particularly relevant in the field of healthcare.

Article 6 of the AI Act identifies two groups of AI systems that are considered *high-risk.* On the one hand, the first group includes products already covered by one of the Union harmonization legislations among those listed in Annex I. In order for these products to be qualified as high-risk AI systems, two conditions shall be met: (a) the AI system should be itself a product (*stand-alone* software) or be «intended to be used as a safety component of a product,» and (b) the conformity assessment of such product under the given harmonized regulation is required to involve a third-party conformity assessment body, i.e., a Notified Body shall be involved in the evaluation of the safety and efficacy of the product, before it being placed on the Union market. For what is of interest in this volume, the list of harmonized legislations in Annex I includes both the MDR and the IVDR and thus, consequently, software-medical devices and in vitro devices.

As a consequence, two conditions should be met in order for software generally implemented and used in healthcare to be considered a *high-risk AI system* under the AI Act. The first condition relates to the qualification

---

[39] European Commission, Guidelines on prohibited artificial intelligence practices, 2024

of the product itself: the software under consideration should be (a1) a stand-alone software qualified as a *medical device* under the MDR, and thus falling within the definition provided for therein as «any (...) software (...) intended by the manufacturer to be used, alone or in combination, for human beings for» one or more medical purposes among those listed by the regulation,[40] or alternatively (a2) a *safety component* of the latter, i.e., «a component of a product or of an AI system which fulfils a safety function for that product or AI system, or the failure or malfunctioning of which endangers the health and safety of persons or property».[41] Moreover, the second condition (b) refers to the type of conformity assessment that the product should undergo, and in particular a Notified Body should be involved in testing the product, verifying its characteristics and releasing the CE Certificate. According to the MDR, this last condition is met when the medical device is classified in class IIa, IIb, or III. If both a and b are cumulatively met, the software would qualify at the same time as a *medical device* under the MDR (of class IIa or higher) and a *high-risk* AI system under the AI Act, provided that it complies with the definition of «AI system» established by the latter (and extensively discussed in the previous pages).

Article 6 then identifies another group of high-risk AI systems, namely those specifically listed in Annex III. In the context of healthcare, reference shall be made to the AI systems intended «to be used by public authorities (or on their behalf) to evaluate the eligibility of natural persons for essential public assistance benefits and services, (...) as well as to grant, reduce, revoke, or reclaim such benefits and services,» and «to evaluate and classify emergency calls by natural persons or to be used to dispatch, or to establish priority in the dispatching of, emergency first response services, including by police, firefighters and medical aid, as well as of emergency healthcare patient triage systems».[42] Notwithstanding the fact that these systems are used for providing healthcare services, broadly defined and considered, they do not usually fall under the definition of *medical device*, because of the lack of a direct inference on the clinical pathway of the patient.[43]

---

[40] The medical purposes that qualify a product as a medical device are: (1) the diagnosis, prevention, monitoring, prediction, prognosis, treatment, or alleviation of disease, (2) the diagnosis, monitoring, treatment, alleviation of, or compensation for, an injury or disability, (3) the investigation, replacement or modification of the anatomy or of a physiological or pathological process or state, or (4) providing information by means of in vitro examination of specimens derived from the human body, including organ, blood, and tissue donations.

[41] Article 3, paragraph 14, AI Act. The same reasoning applies also to software that qualify as in-vitro devices according to the IVDR.

[42] Annex III, point 5, AI Act.

[43] More on the characteristics of a software to be qualified as a medical device

As it appears from the analysis of the mentioned provisions, the European co-legislators decided for both groups of high-risk AI systems to specifically list the cases in which the risk posed by the technology is considered serious enough to raise the need for protecting fundamental rights and, consequently, to legitimize imposing stringent obligations on the provider and the deployers of such system. While the list in Article 6 paragraph 2 refers to *types* of products and consequently to the regulations that govern them, Article 6 paragraph 3 lists *situations* or *environments* in which using an AI system is considered to be posing a high-risk per se. By jointly considering the cases covered by either of the mentioned provisions, it becomes apparent that most AI systems implemented for healthcare will be considered high-risk. However, some of the applications described in Chapter 1 are probably excluded. Reference can be made, for instance, to AI systems used for policymaking, where the technology does not qualify either as a medical device (because of the lack of a direct medical purpose) or as an in-vitro device or as one of the systems listed in Annex III.

4.2.3 The product categories identified by the AI Act—(3) limited-risk AI systems and (4) minimal or no-risk AI systems

Only the first and second groups of AI identified above (prohibited AI systems and High-risk AI systems) are explicitly mentioned in the regulation, while the other two can be deduced from its structure and norms. As a consequence, AI systems with limited, minimal, or no risks are all those software that fall within the definition of artificial intelligence according to the AI Act, but do not meet the requirements of Article 6 to be qualified as high risk.

More specifically, the category of *limited-risk* AI can be inferred from Article 50 AI Act. Indeed, while the vast majority of the norms included therein specifically address and regulate the design, manufacturing, and marketing of high-risk AI systems, article 50 that opens and closes Chapter IV of the AI Act imposes transparency obligations on providers or deployers of specific AI systems, namely those aimed at (3a) interacting with natural persons, (3b) generating synthetic audio, image, video or text content, (3c)

---

on the MDCG 2019-11 Guidance on qualification and classification of software in Regulation (EU) 2017/745—MDR and Regulation (EU) 2017/746—IVDR, October 2019. The guidance was endorsed by the Medical Device Coordination Group (MDCG) established by Article 103 of Regulation (EU) 2017/745. The MDCG is composed of representatives of all Member States and it is chaired by a representative of the European Commission.

generating or manipulating images, audio or video contents constituting a deepfake, or that are (3D) emotion recognition system or a biometric categorization system (when not prohibited by Article 5). Providers and deployers of these AI systems are exempt from complying with most of the provisions of the regulation, but should nonetheless comply with the transparency obligations set forth in Article 50 and discussed extensively below.

Finally, AI systems that do not meet any of the mentioned conditions and thus do not fall in any of the described categories are included among the *minimal risk* or *no risk* AI systems, which the European co-legislator decided not to regulate under the AI Act in accordance with the principle of proportionality. Indeed, the only provision relevant to these systems is Article 95 which encourages the adoption of Codes of Conduct for the voluntary application of the norms set forth therein. As a consequence, providers and deployers of these AI systems may choose to comply with Codes of Conduct enacted under Article 95, but are otherwise generally exempted from the application of the AI Act.

4.3 The room for *maneuver* in the classification established by the AI Act

The top-down architecture adopted by the AI Act establishes a risk pyramid for both AI systems and general-purpose AI systems. This pyramid is meant to be applied equally throughout the European Union among Member States to guarantee legal certainty for the norms' targets and the natural persons whose rights and interests are protected. It also contributes, with a theoretical classification, to the future-proof nature of the regulation itself.

However, this approach raises various issues,[44] in particular related to the choice of implementing the risk-based approach by providing closed and static lists of instances of «risky» AI at various degrees. More specifically, (1) AI systems that pose an unacceptable risk are specifically listed in Article 5, and (2) the instances in which AI systems shall be qualified as high risk are alternatively listed in Annex I (for product regulations) or Annex III (for specific uses). Such an approach to risk-based regulation requires legislative amendments for modifying the regulation and as a consequence may lack the necessary flexibility to adapt to future developments of the technology that may equally pose concrete risks to fundamental rights and value but not be (already) included in the lists provided for by the regulation, possibly

---

[44] Martin Ebers, «Truly risk-based regulation of artificial intelligence—how to implement the EU's AI Act», *op. cit.*; Claudio Novelli *et al. op. cit.*

hindering the aim of the European legislation to establish a *future-proof* regulation on the matter.

To mitigate the rigid approach apparently adopted, the European co-legislators foresaw the possibility that these categories would not be suitable in concrete circumstances or that such lists would need to be expanded in the future. Therefore, the AI Act includes norms for derogating from or amending the imposed regulatory regime.

Indeed, firstly, Article 6 paragraph 4 establishes that «[a] provider who considers that an AI system referred to in Annex III is not high-risk shall document its assessment before that system is placed on the market or put into service.» The instances in which this is possible are when «(a) the AI system is intended to perform a narrow procedural task; (b) the AI system is intended to improve the result of a previously completed human activity; (c) the AI system is intended to detect decision-making patterns or deviations from prior decision-making patterns and is not meant to replace or influence the previously completed human assessment, without proper human review; or (d) the AI system is intended to perform a preparatory task to an assessment relevant for the purposes of the use cases listed in Annex III.»[45] In order to apply the mentioned exemption, the provider shall declare their intention to derogate from the standard qualification as high risk.

Moreover, with reference to high-risk AI systems listed in Annex III, the AI Act empowers the European Commission both to directly modify it on certain grounds,[46] and to add new conditions under which the provider may apply the mentioned exemption or to modify[47] or delete[48] the existing ones, «where there is concrete and reliable evidence of the existence of AI systems that fall under the scope of Annex III, but do not pose a significant risk of harm to the health, safety or fundamental rights of natural persons».[49] All these cases, i.e., when the provider applies the mentioned exemption or the European Commission amends the AI Act, are based on the performance of a *concrete* impact assessment on the AI system to be developed, as opposed to the *a priori* and *theoretical* one already provided for by the AI Act.

As it is apparent, Article 6 Paragraph 4 only partially makes the regulation more flexible, given its nature as an *exception* to the ordinary approach, and any amendments on the part of the European Commission would have to follow the complex procedure of Article 97. Moreover, the European

---

[45] Article 6, paragraph 3, AI Act.
[46] Article 7, AI Act.
[47] Article 6, paragraph 6, AI Act.
[48] Article 6, paragraph 7, AI Act.
[49] Article 6, paragraph 6, AI Act.

co-legislators confirm also for these AI systems downgraded in terms of risks the duty to register in the EU database established by Article 71, as generally applicable for high-risk AI systems of Annex III.

At the same time, Article 112 provides for a system for evaluating and reviewing the need for amendments to the list of prohibited AI systems every four years starting from August 2, 2028.

Quite differently, for high-risk AI systems that are products (or safety components of a product) regulated under one of the Union harmonization legislation listed in Annex I, the AI Act relies more heavily on the evaluation of the level of risk posed by the technology performed according to such legislation, considering that no derogation is possible in this regard. The only exception provided for is a general possibility for the European Commission to «submit appropriate proposals to amend [the AI Act], in particular taking into account developments in technology, the effect of AI systems on health and safety, and on fundamental rights, and in light of the state of progress in the information society».[50]

Only time and the practical application of the norms will tell if the European co-legislators concrete approach to substantiating the risk-based approach will be sufficiently sound and effective.

4.4 How the AI Act regulates high-risk AI systems

4.4.1 Conformity assessment procedures, technical requirements, and the interplay between the AI Act and the MDR

As mentioned, the majority of norms of the AI Act are devoted to regulating the design and marketing of high-risk AI systems, a category in which most of the AI systems implemented in healthcare will be classified. Therefore these provisions are particularly relevant in the field under consideration, as they will be those ensuring the safety and efficacy of the systems deployed for the provision of healthcare. For high-risk AI systems, the AI Act adopts the typical structure[51] of other risk-based regulations composed of 3 phases[52]: (I) *risk assessment and categorization*, provided for by the AI Act it-

---

[50] Article 112, paragraph 10, AI Act.

[51] Tobias Mahler, «Between risk management and proportionality: the risk-based approach in the EU's Artificial Intelligence Act proposal», in Luca Colonna, and Rolf Greenstein (ed. by), *Nordic Yearbook of Law and Informatics 2020–2021: Law in the Era of Artificial Intelligence*, The Swedish Law and Informatics Research Institute 2022, 2022, pp. 245-267.

[52] Martin Ebers, «Truly risk-based regulation of artificial intelligence—how to

self, as discussed above; (II) *impact assessment*, which is one of the mandatory steps of the (III) *risk management*[53] system that the provider of the AI shall have in place according to Article 9. The fundamental aim of this approach is for the entity responsible for the technology to have in place an organizational system able to address and mitigate the concrete risks posed by AI at first at the time of designing the technology and before it being placed on the market, but then also throughout its entire lifecycle.

Being explicitly a risk-based regulation, the main goal of the AI Act is to mitigate possible risks posed by AI systems according to the proportionality principle, avoiding both over- and under-regulating them. To this end, for high-risk AI systems, the rules generally replicate the regulatory structure typical of other product regulations, by establishing the duty to perform a conformity assessment procedure to identify possible risks and prove that the AI system has been designed in compliance with the specific requirements set forth therein to mitigate them, to draw up the declaration of conformity, to affix the CE mark on the system and to conduct post-market activities (article 16). These requirements are developed with the aim of ensuring the trustworthiness of the system, and therefore are related to transparency, the characteristics of the technical documentation, the duty to guarantee human oversight, but also data governance, record-keeping, accuracy, robustness and cybersecurity.

The specific design and development requirements for AI to be compliant with are those included in Chapter III Section 2, and in general require «to ensure that their operation is sufficiently transparent to enable deployers to interpret a system's output and use it appropriately» (Article 12), and of guaranteeing «effectively oversee by natural persons during the period in which they are in use (…) to prevent or minimise the risks to health, safety or fundamental rights that may emerge» with measures that are «commensurate[d] with the risks, level of autonomy and context of use of the high-risk AI system» (Article 14). Moreover, the AI Act requires an adequate level of accuracy, robustness, cybersecurity, and consistent performance of the technology throughout the entire lifecycle of the AI (Article 15). The last provision, in particular, aims at ensuring that the level of safety and efficacy attained before, and in order to, placing the system on the market is preserved through time.

For high-risk AI systems that are medical devices, the software shall also comply with the safety and performance requirements, and the provider shall conduct the conformity assessment procedure, draw up the relevant

---

implement the EU's AI Act», *op. cit.*
[53] Claudio Novelli *et al.*, *op. cit.*

documentation, and perform post-market surveillance activities as set forth in the MDR.

This framework could have led to contrasts between the two regulations or an excessive increase of administrative activities. As a consequence, the AI Act established that in order to «ensure consistency, avoid duplication and minimise additional burdens,» the provider can integrate the necessary testing and reporting processes, information, and documentation required by the AI Act into the documentation and procedure that already exist under the MDR.[54] However, neither regulations precisely determine how to perform such harmonization, and as a consequence, the difficulties in this regard will relate to the duty to correctly interpret the provisions of both regulations and harmonize them in order to design and manufacture a compliant AI system-medical device and draw up and keep up to date the documentation required in both cases.

While the conformity assessment procedure now described may be relatively straightforward for static AI systems, the manufacturer/provider may face some challenges in the case of dynamic systems that learn and continuously change through time and thanks to real-world experiences and data. Indeed, such a procedure aims to identify safety risks posed by the medical device before the product can be placed on the market and in order to mitigate them before making it available for use on the Union market. As a consequence, changes in the system through time may add new risks not envisioned when the assessment was originally conducted and thus potentially endanger patients and users[55]. For these reasons, notified bodies, i.e., the entities entitled to validate the conformity assessment procedure performed by the manufacturer and issue the CE certificate to confirm its safety, usually ask the manufacturer to «'frozen' [the system] at a certain point in the learning process to evaluate [it] at that point.» The consequence of this approach is that anything that is learnt by the AI after being placed on the market will likely affect the conformity of the product to the MDR.[56] Considering the adverse impact that this approach could have on the development of the technology and its innovative nature, new solutions to the issue of dynamic

---

[54] Article 8, paragraph 2, AI Act.

[55] Barry Solaiman, and Mark Bloom, «AI explainability, and safeguarding patient safety in Europe: towards a science-focused regulatory model», in Ivan Glenn Cohen, and Timo Minnssen *et al.*, (ed. by), *The future of medical device regulation*, Cambridge University Press, 2022, pp. 91-102.

[56] Ebers, Martin, «AI robotics in healthcare between the EU medical device regulation and the Artificial Intelligence Act: Gaps and inconsistencies in the protection of patients and care recipients», *Oslo Law Review*, 11 (1), 2024, pp. 1-12.

AI systems will need to be identified, preferably in a dialogue among stake-holders, institutions and notified bodies.[57]

All the mentioned requirements may be said to constitute the concrete application of the precautionary principle, and thus of the assumption that new technologies, and AI is no exception, should not be marketed and used unless «proven safe enough».[58] However, innovating from the standard regu-latory model for the product regulations that are part of the New Legislative Framework, but adopting a standard precautionary practice from overseas, the AI Act introduces for the first time in the European market the concept of *regulatory sandboxes* to which a relatively high number of norms is dedi-cated in various articles in Chapter VI (from 57 to 61 included). These are controlled experimentation environments for the development and pre-mar-keting phase of the development of an AI, which are used with the general aim of ameliorating and testing the compliance of the innovative AI systems with the AI Act and other relevant applicable norms.[59]

However, the approach adopted by the AI Act for high-risk AI sys-tems, which follows the standardized one of the regulations of the New Legislative Framework, has also been criticized for placing too much trust in the provider of the system itself.[60] Indeed, it is the latter that will be responsible for deciding how to concretely implement most of the safety mechanisms provided for by the AI Act, such as the Risk Management System, the Data governance regime, drawing up the correct technical documentation, etc.

4.4.2 Challenges related to the regulation of in-house AI-medical devices

For the manufacturing and marketing of medical devices, the MDR pro-vides for a generally applicable framework and for a special framework in

---

[57] For some proposals in this regard, see Anastasiya Kiseleva, «AI as a medical device: is it enough to ensure performance, transparency and accountability?», *Euro-pean Pharmaceutical Law Review,* 4(1), 2020, pp. 5-16; Ebers, Martin, «AI robotics in healthcare between the EU Medical Device Regulation and the Artificial Intelli-gence Act: gaps and inconsistencies in the protection of patients and care recipients», *op. cit.*; Kerstin Vokinger, Thomas J. Hwang, and Aaron S. Kesselheim, «Lifecycle regulation and evaluation of artificial intelligence and machine learning-based medi-cal devices», in Ivan Glenn Cohen *et al.* (ed. by), *The future of medical device regulation,* Cambridge University Press, 2022, pp. 13–21.

[58] Margot E. Kaminski, «Regulating the Risks of AI», *op. cit.*

[59] Recital 139, AI Act.

[60] Michael Veale, and Frederik Zuiderveen Borgesius, «Demystifying the draft EU Artificial Intelligence Act», *Computer Law Review International*, 22(4), 20021, pp. 97-112.

Article 5 paragraph 5 specifically dedicated to *in-house* medical devices, which are devices, including software, manufactured and used exclusively by and within a specific healthcare institution. For these products, under certain strict conditions, the regulation establishes multiple derogations to the generally applicable principles and obligations. In particular, the healthcare institution has the possibility to choose the lighter regulatory regime enshrined in Article 5 paragraph 5, which provides for a duty to comply with the technical requirements set forth by the MDR for the manufacturing of the product as well as some of the other general obligations established therein, such as having a quality management system in place and collaborate with the competent authority upon request. However, on the contrary, the healthcare institution is not obliged to comply with most of the other generally established obligations, such as affixing the CE marking on the device, drawing up the ordinary documentation and having a post-market surveillance system in place.

The requirements for the application of this regulatory simplification are (1) *subjective* elements, answering the question of who can apply the mentioned regulatory regime; (2) *objective* elements, i.e., for which products this regime can be applied; and (3) other conditions to be observed, in particular which activities shall be carried out by the entity that manufactures and uses the device, and more generally which conditions must be observed to be compliant with the MDR. In this regard, insights and guidance on the interpretation and application of Article 5 paragraph 5 MDR is provided for by the MDCG 2023-1 Guidance on the health institution exemption under Article 5(5) of Regulation (EU) 2017/745 and Regulation (EU) 2017/746, a non-legally binding (but highly considered) interpretative instrument issued by the Medical Device Coordination Group.

1. Subjectively, Article 5 paragraph 5 MDR can only be applied by a health institution established in the EU, defined by the MDR as «an organisation whose main purpose is the care or process of patients or the promotion of public health».[61] From this notion, Recital 30 explicitly excludes «hospitals and institutions, such as laboratories and public health institutes that support the health system and/or meet the needs of patients, but *which are not directly involved in the process or treatment of patients*» (emphasis added) and also «businesses whose main stated aims are related to health and healthy lifestyles, e.g. gyms, spas, wellness and fitness centres.»

---

[61] Article 2, paragraph 36, AI Act.

2. From an objective point of view, however, the exception provided for in-house medical devices applies only to those devices that cumulatively meet two conditions, namely that (2a) are devices that are not manufactured on an industrial scale and are used exclusively in healthcare institutions within the European Union, and (2b) they are suitable to meet specific needs of the target patient group, which cannot be met or cannot be met with results of the same level by an equivalent device available on the market. To comply with this second requirement, the healthcare institution must first identify the target patient group and its needs, then verify that there are no equivalent devices available on the market, and after having applied the described exception constantly update this assessment.

The verification of the non-existence of a medical device equivalent to the in-house device on the market must be carried out with reference to the national market of the Member State where the healthcare institution is located and aims to ascertain that no technology on the market achieves the same impact on the patient's care and healing journey, with equivalent technology and overlapping technical and clinical risks, taking into the technical, biological, and clinical characteristics of the compared devices. Should the institution find during these updating activities that an equivalent device has been placed on the market, it will have to proceed with the discontinuation of the in-house medical device and the adoption of the available CE-marked one.

The applicability of the AI Act to this type of medical devices may not be straightforward. Indeed, in-house devices are products with the same intended purpose of regular medical devices, but which follow a simplified regulatory pathway because of special needs of a target group of patients that could not otherwise be addressed promptly if the health institution had to undergo the traditional conformity assessment procedure or not at all with devices on the Union market.

However, *in-house* devices may well be software and AI systems within the meaning of the regulation and may perform complex tasks, which consequently may pose significant risks to patients. As such, and adopting the precautionary and proportionality principles, *in-house* devices not only should be included within the material scope of the AI Act, but should also, in principle, be qualified as *high-risk* because of the possible consequences that may arise from their use. However, these devices do not undergo a conformity assessment procedure involving a Notified Body, and thus they would not be qualified as *high-risk* AI systems, unless included in one of the alternatives in Annex III.

In this regard, further information or interpretative documents will be needed to ascertain whether the European co-legislators intended to provide a lighter regime for *in-house* AI medical devices, mimicking the approach adopted by the MDR or did not take these devices into consideration from the outset.

4.4.3 Other relevant provisions for high-risk AI systems: transparency (Article 13), human oversight (Article 14), and fundamental rights impact assessment (Article 27)

Apart from the requirements established to ensure that the high-risk AI system is safe and effective, the AI Act provides for other norms to guide the design, and consequently use, of the technology. Among them, (I) Article 13 on transparency requirements, (II) Article 14 on human oversight measures; and (III) Article 27 fundamental rights impact assessment.

1. First of all, the AI Act gives legal form to the principle of transparency in Article 13 that establishes explainability requirements for high-risk AI systems. This provision has been qualified as both *user-empowering* and *compliance-oriented*, because «on the one hand, it serves to enable users of the AI system to use it correctly; on the other hand, it helps to verify the adequacy of the system to the many obligations set by the AIA, ultimately contributing to achieving compliance».[62] Indeed, paragraph 1 provides that high-risk AI systems «shall be designed and developed in such a way as to ensure that their operation is sufficiently transparent to enable deployers to interpret a system's output and use it appropriately» (*user-empowering explainability provision*). To this end, the provider should market the system accompanying it with instructions for use, which shall specify not only information about herself but also some technical characteristics of the technology, such as the characteristics, capabilities, and limitations of performance, level of accuracy, specifications of the intended use, and on the characteristics of input data, as well as the human oversight measures adopted.[63] This detailed provision led some early commentators claiming that «EU AI regulations won't permit an opaque system»[64], and that according to the newly enacted

---

[62] Francesco Sovrano *et al.*, «Metrics, explainability and the European AI Act Proposal», *Multidisciplinary Scientific Journal*, 5, 2022, pp. 126-138.
[63] Article 13, paragraph 3, AI Act.
[64] https://www.forbes.com/sites/glenngow/2021/10/10/the-eu-is-regulating-your-aifive-ways-to-prepare-now/.

norm deep learning must be banned. Quite on the contrary, Recital 72 of the AI Act provides clarity on the purpose of Article 13, by claiming that transparency is necessary to address the potential opacity and excessive complexity of the system for natural persons. The solution found by the European co-legislators in this regard was to establish that specific documentation and instructions for use should accompany the AI system. Consequently, the chosen approach was to establish information requirements in the documents accompanying the AI system to guarantee that excessive opacity or unintelligibility is avoided. What the AI Act does not prescribe are «specific transparent-by-design models or [the] mandatory use of XAI tools,» which however can be freely chosen as an additional measure by the provider.[65] Moreover, at the same time, the second part of Article 13, paragraph 1 establishes that «[a]n appropriate type and degree of transparency shall be ensured with a view to achieving compliance with the relevant obligations of the provider and deployer set out in Section 3» (*compliance-oriented explainability provision*). Finally, it has been highlighted that both sides of the explainability provision in the AI Act are enshrined in Article 26, paragraph 5, where it is established that deployers shall be able to «monitor the operation of the high-risk AI system on the basis of the instructions of use,»[66] which however requires that first of all the user shall have the necessary information to understand the reasoning and outputs of the AI system, that the system itself had been designed in an intelligible way and that all the necessary documents have been provided to the user. In asking providers to comply with a certain level of transparency, the AI Act integrates the MDR, which lacks specific and explicit provisions in this regard, even though it had been argued that «a certain level of transparency is a prerequisite» to comply with the other product requirements set forth therein, namely demonstrating the safety and performance of the device before placing it on the market, as well as providing sufficient and adequate information on the functioning of the product to the user.[67]

---

[65] Cecilia Panigutti *et al.*, «The role of explainable AI in the context of the AI Act», in *Proceedings of the 2023 ACM Conference on Fairness, Accountability, and Transparency (FAccT '23)*, Association for Computing Machinery, 2023, pp. 1139-1150.

[66] Francesco Sovrano *et al.*, *op. cit.*

[67] Anastasiya Kiseleva, Dimitris Kotzinos, and Paul De Hert, «Transparency of AI in healthcare as a multilayered system of accountabilities: between legal requirements and technical limitations», *Frontiers in Artificial Intelligence*, 5, 2022, pp. 1-21; Ebers, Martin, «AI robotics in healthcare between the EU medical device regulation and the Artificial Intelligence Act: gaps and inconsistencies in the protection of patients and care recipients», *op. cit.*

2. Moreover, coherently with the principle of respect for autonomy of human agents, Article 14 of the AI Act requires any high-risk AI system to be designed «in such a way, including with appropriate human-machine interface tools, that they can be effectively overseen by natural persons during the period in which they are in use,» with the aim of preventing or minimizing «the risks to health, safety or fundamental rights that may emerge when [the] high-risk AI system is used in accordance with its intended purpose or under conditions of reasonably foreseeable misuse» and taking into consideration «the level of autonomy and context of use of the high-risk AI system».[68] The identified measures should enable the deployer:

I.   to properly understand the relevant capacities and limitations of the high-risk AI system and be able to duly monitor its operation (...);

II. to remain aware of the possible tendency of automatically relying or over-relying on the output produced by a high-risk AI system (automation bias) (...);

III. to correctly interpret the high-risk AI system's output (...);

IV. to decide, in any particular situation, not to use the high-risk AI system or—to otherwise disregard, override, or reverse the output of the high-risk AI system;

V.   to intervene in the operation of the high-risk AI system or interrupt the system through a «stop» button or a similar procedure (...).

On the matter, Recital 73 clarifies that «such measures should guarantee that the system is subject to in-built operational constraints that cannot be overridden by the system itself and is responsive to the human operator, and that the natural persons to whom human oversight has been assigned have the necessary competence, training and authority to carry out that role».[69] As a consequence, it appears that the AI Act requires human oversights measures to be built into the AI system before it is placed on the it is placed on the European Union market and therefore made available for use, when technically feasible.[70]

---

[68] Article 14, AI Act.

[69] For concrete models for designing AI systems in compliance with this principle, see for instance Doris Aschenbrenner *et al.*, «Research interpretation of Article 14 of the EU AI Act: human in command in manufacturing», in Matthias Thürer *et al.* (ed. by), *Advances in production management systems. Production management systems for volatile, uncertain, complex, and ambiguous environments*, Springer, 2024, pp. 226-239.

[70] Claes G. Granmar, «AI-based decision-making and the human oversight requirement under the AI Act», in Eduardo Gill-Pedro, and Andreas Moberg (ed. by), *YSEC Yearbook of socio-economic constitutions 2023*, Springer, 2023, pp. 181-212.

3. Finally, Article 27 of the AI Act establishes a duty to perform an «assessment of the impact on fundamental rights that the use of such system may produce» (so-called Fundamental Rights Impact Assessment—FRIA), taking into consideration in particular the concrete circumstances of the use of the high-risk AI system, in terms of characteristics of the deployers, time, and frequency of use, categories of natural persons possibly affected by it, risks or harm likely to occur, and measures envisioned to mitigate them. The duty to conduct a FRIA is not applicable to any deployers or any high-risk AI systems, but exclusively to:

- bodies governed by public law or private entities providing public services if they are deployers of a system listed in Annex III, with the exclusion of those in point 2—thus for instance a FRIA shall be performed when implemented for determining access to and enjoyment of essential private services and essential public services and benefits, including healthcare services;

- deployers of high-risk AI systems intended to be used to evaluate the creditworthiness of natural persons or establish their credit score, with the exception of AI systems used for the purpose of detecting financial fraud, or to be used for risk assessment and pricing in relation to natural persons in the case of life and health insurance. This assessment resembles to some extent the Data Protection Impact Assessment required by the GDPR and further described below. Indeed, the same AI Act recognizes this parallelism, but establishes in paragraph 4 of Article 27 that «[i]f any of the obligations laid down in this Article is already met through the data protection impact assessment conducted pursuant to Article 35 of Regulation (EU) 2016/679 or Article 27 of Directive (EU) 2016/680, the fundamental rights impact assessment referred to in paragraph 1 of this Article shall complement that data protection impact assessment.» However, the two instruments differ in the object of the assessment, being the evaluation of potential infringements of the right to data protection in the DPIA and of any fundamental right or freedom in the FRIA. At the same time, this instrument had been adapted by the European Parliament to AI from the Human Rights Impact Assessment (HRIA), a policy tool established at the international level usually as part of ex-post due diligence procedures performed by private and business companies.[71] HRIA is the

---

[71] Alessandro Mantelero, «The fundamental rights impact assessment (FRIA) in the AI Act: roots, legal obligations and key elements for a model template», *Computer, Law & Security Review*, 54, 2024, pp. 1-18.

«process for identifying, understanding, assessing and addressing the adverse impacts of projects and activities on the enjoyment of human rights in relation to potentially affected rights-holders».[72]

The main characteristics of an FRIA performed according to the AI Act are that it adopts an ex-ante approach, being conducted before the AI system is used; it has «a rights-based focus and a circular iterative structure that follows the product/service throughout its lifecycle»; and finally it has «an expert-based nature,» in the sense that it should be conducted by experts in the field.[73] Under a more practical point of view, the FRIA to be conducted resembles in the structure the conformity assessment procedure required by the regulation but quite differently from the latter is highly context-dependent and not completely possible to standardize.

4.5 How the AI Act regulates any type of AI systems

As mentioned, most of the provisions of the AI Act are exclusively applicable to high-risk AI systems, especially those related to the requirements for the design of the technology. Indeed, these systems do not have to undergo a conformity assessment procedure, but shall merely comply with the (admittedly limited) general provisions that apply also to them. Indeed, apart from the norms specifically dedicated to general-purpose AI systems and models, the AI Act provides for some obligations imposed on the provider or deployer of limited and minimal risk AI systems as well, thus irrespective of their risk classification and intended purpose. The following deserve special attention: Article 50 on transparency obligations for limited-risk AI systems, and Article 4 on AI literacy.

In particular, for *limited-risk* AI systems, the AI Act mandates in Article 50 that the natural person who interacts with the system or comes into contact with the contents generated by the AI be aware and adequately informed of the existence of the AI behind the technological curtains and its role and involvement in the creation and development of such content. Once again, the AI Act aims at ensuring that the ethical principles of explicability, human autonomy, and especially transparency are adequately complied with also in the case of AI systems that were not classified as high-risk. This way,

---

[72] United Nations—Human Rights Council, «Guiding principles on business and human rights: implementing the United Nations 'protect, respect and remedy' framework», Resolution 17/4 of 16 June 2011; Alessandro Mantelero, *op.cit.*
[73] Alessandro Mantelero, *op.cit.*

the European co-legislators attempted at permeating every level of AI system design with the mentioned principles. However, this provision and the related obligations are applicable only to the AI systems classified as limited risk, and therefore to those listed in Article 50 itself.

Finally, in order to address issues related to the ethical principle of equity, the AI Act at its very beginning provides for a general duty for providers and deployers of any kind of AI system (thus including any system from *high-risk* to *no risk*) to «take measures to ensure, to their best extent, a sufficient level of AI literacy of their staff and other persons dealing with the operation and use of AI systems on their behalf, taking into account their technical knowledge, experience, education and training and the context the AI systems are to be used in, and considering the persons or groups of persons on whim the AI systems are to be used».[74]

4.6 AI Act and the interplay with the legislative framework for data processing

It goes without saying that data are essential in the context of AI, especially in healthcare, both to train the model adequately and to reach its intended use concretely. Basically, AI would not be possible without data. As a consequence, the very existence of AI systems strictly depends on the quantity and quality of data available that can be collected, processed, and stored for the design, training, deployment, and improvement of the technology throughout the entire life cycle of these products.

In the European Union, the legislative framework applicable to the processing of personal data consists of various components, the most fundamental of which is the already commented Regulation 2016/679 (GDPR), and others are progressively being added within the framework of the European Digital Strategy, first published in 2020. As notorious, with the mentioned strategy, the European Union aims to foster data-driven innovations, by incrementing the possibilities for data sharing, and developing a single market for data, while at the same time providing data subjects with more instruments to (re)gain control over their data and, as a consequence, safeguarding their fundamental rights.[75] Among the multiple regulations envisioned in the strategy, the most relevant in the context of AI are Regulation 2022/868,

---

[74] Article 4, AI Act.
[75] Communication From the Commission to the European Parliament, the Council, The European Economic and Social Committee and the Committee of the Regions, *A European strategy for data,* 2020.

the so-called Data Governance Act (DGA), the Regulation 2025/327 on the European Health Data Space (EHDS), and Regulation 2023/2854, the so-called Data Act (DA).

The following pages will be devoted to discussing the interplay between the AI Act and the GDPR, as well as the innovations established by the EHDS and the DGA for the processing of personal and non-personal data also for the purposes of training AI systems. Differently, the goal of the DA, «probably the most innovative data-related legal regime deployed by the EU» is to lay down «harmonized rules on fair access to and use of data» in the internal market and thus somehow impacts on the second phase of the AI systems life-cycle, i.e. their use.

Indeed, the DA stems from the idea that the generation of data by connected products (or related services) «is the result of the actions of at least two actors»: the designer/manufacturer of the product, and the user of that product. As a consequence, the new set of norms aims at addressing the questions of «fairness in the digital economy» by dealing with the issues of the assignment of rights regarding the access to and the use of these data among all the stakeholders, entities and people involved. The architecture established by the regulation in particular grants the user of a connected product or related service with some rights against the data holder, i.e. who has control over the data processed by the product and thus usually the manufacturer/provider. Among these rights, there is that of accessing (by default or upon request) the data and sharing them with third parties (data portability). Despite the complex set of provisions set forth therein, it appears evident that this regulation is meant to have a significant impact in the field of AI, in terms of obligations to be complied with in order to make the AI systems accessible by design, data sharing procedures to have in place, and further transparency requirements to take into consideration.

4.6.1 AI and the GDPR

For the development and deployment of AI systems in healthcare «data protection is a critical matter,» and that between the AI Act and the GDPR has been defined as a «complex connection.»[76] Indeed, the vast amount of high-quality data needed, usually genetic data and data concerning the health of patients, calls for special attention on their processing, to ensure

---

[76] Janos Meszaros, Jusaku Minari, and Isabelle Huys, «The future regulation of artificial intelligence systems in healthcare services and medical research in the European Union», *Frontiers in Genetics*, 13, 2022, pp. 1-10.

that a fair balance is struck between competing interests and, thus, that the fundamental rights of patients are adequately protected.

In Europe, the primary regulation to comply with in this regard is the GDPR. The GDPR is a principle-based regulation,[77] provided that Article 5 lays down the principles that shall be complied with in any processing of personal data: (a) lawfulness, fairness, and transparency; (b) purpose limitation; (c) data minimization; (d) accuracy; (e) storage limitation; (f) integrity and confidentiality. The applicability of the mentioned principles in the context of AI in healthcare may be challenged by the very nature of AI and some of its possible applications.

4.6.1.1 The principle of lawfulness, fairness, and transparency—considerations on the right to an explanation and the right not to be subject to a decision based solely on automated process in the context of AI systems in healthcare

(a) First of all, according to the principle of lawfulness, fairness, and transparency, data should be «processed lawfully, fairly and in a transparent manner in relation to the data subject».[78] On a general level, it seems that black-box AI systems may not only be illegitimate according to the specific provisions applicable to them as products as mentioned (and frequently violate ethical principles) but also incompatible with the principles of transparency, fairness, and accountability established by the GDPR.[79] Indeed, not being able to understand the reasoning or the output of an algorithm also entails the impossibility of identifying specifically *which* data are processed and *how*, and thus of informing the data subject of the characteristics of the processing and *who* is responsible for the correctness of such processing. Precisely the request for transparency in the GDPR led many commentators to debate the possibility of finding an alleged *right to an explanation* of the decision taken by an algorithm in the GDPR, enshrined in Articles 13, 14 and 22.[80] If recognized, this right

---

[77] Maria Tzanou, *Health data privacy under the GDPR: Big data challenges and regulatory responses*, Routledge, 2023; Chris Jay Hoofnagle, Bart van der Sloot, and Frederik Z. Borgesius, «The European union general data protection regulation: what it is and what it means», *Information & Communications Technology Law*, 28(1), 2019, pp. 65-98.

[78] Article 5 paragraph 1, GDPR.

[79] Mirko Forti, «The deployment of artificial intelligence tools in the health sector: privacy concerns and regulatory answers within the GDPR», *European Journal of Legal Studies*, 13(1), 2021, pp. 29-44.

[80] Bryce Goodman, and Seth Flaxman, «European Union regulations on algorithmic decision-making and a 'right to explanation», *AI Magazine*, 38(3), 2017, pp. 50-57; Sandra Wachter, Brent Mittelstadt, and Luciano Florid, «Why a right to explanation of automated decision-making does not exist in the General Data Protec-

would widen the scope of the transparency requirements provided for in the AI Act by imposing on the provider a higher threshold for compliance. After all, according to the GDPR the data controller, i.e., the «natural or legal person (...) [which] determines the purposes and means of the processing of personal data»[81] and therefore frequently either the provider or the deployer in the case of AI in healthcare, is under the obligation to provide «meaningful information about the logic involved» in an automated decision.[82] This information is moreover critical for the exercise of most of the other rights provided for by the GDPR, especially that of Article 22, i.e., the right of the data subject not to be subject to a decision based solely on automated processing, including profiling.[83] It goes without saying that in order to have the possibility of opposing such processing, one should first of all have been informed of its existence. Coherently with the ban in Article 5 AI Act, Article 22 GDPR establishes that «[t]he data subject shall have the right not to be subject to a decision based solely on automated processing, including profiling, which produces legal effects concerning him or her or similarly significantly affects him or her,» and is therefore aligned with the principle of respect for human autonomy.[84] As a consequence, Article 22 GDPR is likely to be frequently applicable in the case of the implementation of AI systems in the healthcare sector, especially considering that, according to the jurisprudence of the CJEU, the term «decision» should be interpreted broadly, to include any processing of personal data that produces legal effects for the data subject or significantly affects her.[85]

---

tion Regulation», *International Data Privacy Law*, 7(2), 2021, pp. 76-99; Andrew D Selbst, and Julia Powles, «Meaningful information and the right to explanation», *International Data Privacy Law*, 7(4), 2017, pp. 233-242; Lilian Edwards, and Michael Veale, «Slave to the algorithm? Why a 'right to an explanation' is probably not the remedy you are looking for», *Duke Law & Technology Review*, 16(1), 2017, pp. 1-65; Margot E Kaminski, «The right to explanation, explained», *Berkley Technology Law Journal*, 34(1), 2019, pp. 189-218. More precisely, of a negative opinion: Andrew D Selbst, and Julia Powles, *op. cit.*; Bryce Goodman, and Seth Flaxman, *op. cit.* On the contrary, give a positive answer: Maja Brkan, and Grégory Bonnet, «Legal and Technical Feasibility of the GDPR's Quest for Explanation of Algorithmic Decisions: of Black Boxes, White Boxes and Fata Morganas», *European Journal of Risk Regulations*, 11(1), 2020, pp. 18-50; Sandra Wachter, Brent Mittelstadt, and Luciano Floridi, *op. cit.*

[81] Article 4 paragraph 7, GDPR.

[82] Article 13 paragraph 2 letter f, Article 14 paragraph 2 letter g, Article 15 paragraph 1 letter h, GDPR.

[83] Cecilia Panigutti *et al.*, *op. cit.*

[84] Sabine Salloch, and Andreas Eriksen, «What are humans doing in the loop? Co-reasoning and practical judgment when using machine learning-driven decision aids», *The American Journal of Bioethics*, 24(9), 2024, pp. 67-78.

[85] Judgement in SCHUFA Holding, Case C-634/21, para.44. For an extensive analysis of the case, see Claes G. Granmar, *op. cit.*

However, paragraph 4 of the same article provides for exemptions, by affirming that special categories of data, thus including genetic and health-related data, should not be used for solely automated decisions unless the data subject provided explicit consent or the processing is necessary for reasons of substantial public interest, on the basis of EU or Member State law which is proportionate to the aim pursued. Both instances are particularly relevant in the context of healthcare.

Independently on the actual recognition to data subjects of a right to an explanation by the GDPR, the European co-legislators decided to include a mild version of it in Article 86 of the AI Act, but only for certain types of AI systems. Indeed, Article 86 grants the right to «obtain from the deployer clear and meaningful explanations of the role of the AI system in the decision-making procedure and the main elements of the decision taken» to any person who has been subject to a decision made by the deployer based on the output from a high-risk AI system listed in Annex III (such as those implemented for granting access to healthcare services), with the exception of point 2, and who claims to have suffered adverse consequences for their health, safety or fundamental rights. This way, the AI Act codifies the possibility for a person affected by a high-risk AI system to, albeit in specific circumstances considered to pose particular risks.

At the same time, the principle of fairness requires the output of an AI system being *fair*, in the sense of not providing discriminatory results, which depend to a large extent on the dataset on which they have been trained and the other elements extensively analyzed above.

Moreover, black-box AI systems may also be not reconcilable with consent as the legal basis for any processing performed by them, in particular considering the GDPR requirements for valid consent.[86] Indeed, it is hardly conceivable that consent can be both adequately informed and specific if the very functioning of the AI system that will process personal data is not fully intelligible or explainable.

4.6.1.2 The principle of purpose limitation and data minimization in the development of AI systems in healthcare

Furthermore, the principle of purpose limitation is sometimes addressed as the «cornerstone of data protection, and a prerequisite for most other fundamental requirements»[87] and it requires data to be collected only for a spec-

---

[86] Mirko Forti, *op. cit.*
[87] Cécile De Terwangne, «Article 5 Principles Relating to Processing of Personal Data», in Christopher Kuner *et al.*, (ed. by), *The EU General Data Protection Regula-*

ified and legitimate purpose and not further processed in a manner that is incompatible with those purposes. Therefore, on the one hand, this principle may potentially contrast with the idea that when it comes to deciding the level of sufficient data for training an AI system developers may be tempted to believe that «the more, the merrier.» On the other, according to the principle of purpose limitation conceptualized in the way explained above the so-called *secondary use* of personal data is generally prohibited, unless specific exemptions apply. However, the vast amount of health and genetic data collected by healthcare professionals and facilities during and for the provision of healthcare (thus for a specific *primary purpose*) would be invaluable elements for research purposes, and in particular for the development of new AI systems, but this is only possible if one of the exemptions provided for in the GDPR is applicable in a given concrete situation. It should thus not come as a surprise that in this field the GDPR is frequently perceived as an obstacle.

Moreover, it remains to be seen how the principle of data minimization and the right to be forgotten[88] will be operationalized in the context of AI. On the one hand, AI systems frequently have the computational capacity to process high quantity of data (including personal data), which are usually also required to adequately train such a technology. These data, as mentioned, may show the characteristics of big data (or big health data),[89] and are necessary to ensure that the AI systems have been adequately tested to be safe, effective, and non-discriminatory in their outcomes. As a consequence, possible contrasts with the principle of data minimization may occur because the nature of big data and the AI systems entails collecting as much data as possible, while on the contrary, the principle of data minimization requires to do so as little as possible. As it has been highlighted, this «dichotomy is hard to reconcile».[90] On the other, the right to be forgotten was formulated by the CJEU in the judgment *Google Spain*[91] and is applicable in multiple circumstances, such as whenever the data subject withdraws a consent previously provided or the data are no longer needed for the original purpose. However, any input data provided to an AI by the provider in order to train

---

tion (GDPR): A commentary, Oxford Academic, 2020, pp. 309–320.

[88] Article 17, GDPR.

[89] Maria Tzanou, *op. cit.*

[90] Ivana Bartoletti, «AI in healthcare: ethical and privacy challenges», in David Riaño, Szymon Wilk, and Annette ten Teije (ed. by), *Artificial Intelligence in Medicine, AIME 2019. Lecture Notes in Computer Science, vol 11526*, Springer, 2019, pp. 7-12.

[91] Case C.131/12, *Google Spain SL,Google Inc. v. Agencia Espanola de Protecciòn de Datos, Mario Costeja Gonzàlez (2014).*

it or by a healthcare practitioner as input becomes part of the algorithm itself because they are processed by the technology in order to increase its knowledge and reach its intended purpose.

4.6.1.3 Data protection impact assessment of AI systems in healthcare

According to Article 35 paragraph 1 GDPR, prior to the processing of personal data, the controller may have the duty to carry out an assessment of the impact of the envisaged processing on the protection of personal data taking into account the «nature, scope, context and purposes of the processing»[92] itself (so-called data protection impact assessment—DPIA). Such an assessment is not due for any processing, but only for those that are likely «to result in a high risk to the rights and freedoms of natural persons,» especially when new technologies are implemented.[93] To determine the extent of the applicability of Article 35 GDPR to the context of AI systems in healthcare, two scenarios should be distinguished: When AI systems are classified by the AI Act as high-risks, and when they are not (limited or minimal risks AI systems).

In the first case, it is pretty straightforward to envision that a DPIA will be compulsory in cases of high-risk AI, considering that these systems are hardly not qualifiable as «new technology» likely «to result in a high risk» to natural persons, as AI is usually very innovative in nature and the notion of «high risk» for the purpose of applying Article 35 GDPR resembles that of «high-risk» for classificatory purposes in the AI Act.[94] Adopting the same line of reasoning, a DPIA will not be necessary in the second scenario, and thus when AI systems exhibit low or minimal risks.

4.6.2 AI and regulations of the European Strategy for Data

It is by now well understood that training, validating and using AI systems necessitate the availability of vast quantities of data, both personal and non personal, and various regulatory efforts have over time been undertaken to address this need. Norms have indeed been enacted to attempt to strike a fair balance between, on the one hand, encouraging the sharing and reuse of the data generated in the data-driven society, and, on the other hand, ensuring an adequate level of protection for the fundamental rights of the individuals for various reasons involved and public interests.

---

[92] Article 35 paragraph 1, GDPR.
[93] Article 35 paragraph 1, GDPR.
[94] Claes G. Granmar, *op. cit.*

Within this landscape of the new legislative initiatives, ultimately falling within the scope of the European Strategy for Data, the EHDS holds a position of primary importance. It is a health-sector specific regulatory proposal explicitly developed for electronic health data and the first of the envisioned domain-specific common European spaces. It builds upon other horizontal relevant regulations, and in particular on the GDPR and the DGA, and regulates exclusively the processing of electronic health data, both personal and non-personal (Art. 2(2)(c)).

The goal of the EHDS, as envisioned in Recital 1, is twofold. On the one hand, the EHDS aims at improving «natural persons' access to and control over their personal electronic health data in the context of healthcare» and on the other at better achieving «other purposes involving the use of electronic health data in the healthcare and care sectors that would benefit society, such as research, innovation, policymaking, health threats preparedness and response, including preventing and addressing future pandemics, patient safety, personalised medicine, official statistics or regulatory activities» as well as improving «the functioning of the internal market by laying down a uniform legal and technical framework in particular for the development, marketing and use of electronic health record systems [...] in conformity with Union values.» As such, the EHDS «will be a key element in the creation of a strong and resilient European Health Union.» More concretely, it aims at establishing rules for processing electronic health data for primary purposes, i.e. for the concrete provision of healthcare, and secondary purposes, i.e. to support among others health research and innovation and the development of personalised medicine. All this while nonetheless ensuring an adequate level of protection for the rights of the data subjects to whom the data refer and of control over the use of their data. Precisely, the norms devoted to regulating the secondary processing under the EHDS are particularly relevant in the context of AI.

Echoing the theoretical framework initially set forth by the GDPR, though subsequently (at least partially) undermined in its practical implementation by the transposition of its provisions at the national level and by the (at times) restrictive and conservative national interpretations of its provisions, the EHDS aims to establish a space genuinely oriented toward balancing the two interests: that of society in using these data for public purposes and of data subjects in maintaining an adequate level of control over the use of information referred to them, including the most sensitive ones concerning health (i.e. in the words of the GDPR «special categories of personal data»).

The regulatory approach (re)established by the EHDS and theoretically provided for by the GDPR is coherent with the interpretation of the right to data protection as envisioned in Art. 8 of the EU Charter of Fundamental Rights that sees in paragraph 2 the active side of the mentioned right,[95] and thus interprets it as including both the concepts of informational privacy, i.e. managing one's personal data and controlling their processing and use through time, and informational autonomy, i.e. the right to informational self-determination.[96] The latter ensures that data subjects have the power to control the disclosure, use and circulation of their data, and have the right to self-determine in relation to the use of their personal information. The relationship between a person and their personal data may therefore be explained in terms of «person-information-circulation-control» and not as «person-information-secrecy or privacy» anymore.[97]

To these ends, the GDPR envisions a set of various legal bases for the processing of personal data, including but not limited to informed consent, and a bundle of other rights of the data subject such as the right to access the data, to withdraw consent, to ask for a rectification, etc.[98] The EHDS builds upon this framework and provides for rules aimed at promoting the processing of electronic health data for purposes chosen by the regulation itself as being of sufficient public interest and bringing great societal benefit, while at the same time further reinforces the means and instruments available to the data subject for effectively exercise their control.

In particular, among the secondary purposes for which electronic personal data can be processed the EHDS includes «scientific research related to health or care sectors that contributes to public health or health technology assessments, or ensures high levels of quality and safety of healthcare, of medicinal products or of medical devices, with the aim of benefiting end-users, such as patients, health professionals and health administrators, including: (ii) training, testing and evaluation of algorithms, including in medical devices, in vitro diagnostic medical devices, AI systems and digital health applications». (Art. 53 para. 1 lett. e). The electronic data that shall be shared, and thus can be used, to this end are those included in Art. 51, and as a matter of example aggregated data on healthcare needs, data on pathogens that impact human health, healthcare-related administrative data, but also human genomic data, data from wellness applications and medical devices,

---

[95] Matteo Macillotti, «Informed Consent in the Biobanking Context», *European Journal of Health Law*, 19(3), 2012, pp. 271-288.

[96] Maria Tzanou, *op. cit.*

[97] Stefano Rodotà, *Tecnologie e Diritti, op. cit.*

[98] Simona Fanni, *La donazione dei materiali biologici umani ai fini della ricerca biomedica nell'ambito del diritto internazionale e del diritto dell'Unione Europea*, Napoli, Editoriale Scientifica, 2016.

health data from biobanks and associated databases or data from clinical trials, clinical studies, and clinical investigations.

Simply put, the mechanism envisioned by the EHDS entails the involvement of entities qualified as health data access bodies, appointed by Member States and that should comply with specific obligations in terms of level of competence, transparency and independency, to whom data for secondary use should be made available by the data holders, i.e. whoever has the data to be made available, and that in turn may grant access to such data to the data users that requested it.

On paper, the EHDS has the potential for providing high quality and quantity personal and non-personal data to be processed also for the purposes of training and validating AI systems in a controlled environment in which the health data access bodies work as guarantors and the rights of the data subjects are strengthened and operationalised.

Additionally, another potential new source of data to be used for the same purpose is expected to be the data altruism mechanism envisioned by the DGA. This regulation aims to provide a framework for enhancing trust in voluntary data sharing by promoting the reuse of publicly held data, increasing trust in newly introduced data intermediation services, and encouraging the sharing of data for altruistic purposes. As for the latter in particular, the DGA enables the possibility for voluntary sharing data, so called data altruism, on the basis of either the consent of the data subject or permissions of data holders for processing personal and non-personal data respectively «without seeking or receiving a reward that goes beyond compensation [...] for objectives of general interest as provided for in national law, where applicable, such as healthcare, [...] or scientific research purposes in the general interest» (Art. 2 para. 16) To enable this peculiar sharing of data, the DGA provides that the data shared through this modality are collected by entities qualified as data altruism organisations, that should be included in national registries upon complying with specific requirements, such as carrying out data altruism activities, operating on a not-for-profit basis, be legally independent, etc. (Art. 18) and transparency obligations related to their activity. (Art. 20) The data altruism organisations are then those responsible for making these data available for the purposes for which they were shared and safeguarding the rights and interests of the data subjects or the data holders, respectively. The provisions of the DGA related to the data altruism mechanism have yet to be concretely applied, and therefore their impact, especially in the context of scientific research and healthcare, remains to be seen. However, theoretically, the framework so envisioned may be of great importance for the development and training of AI systems, if such processing is considered to be among the permitted ones according to the DGA.

Together, the GDPR, the DGA and the EHDS contribute to the new «scientific research regime 2.0», i.e. the development of a new set of norms applicable to the processing of personal (but also, in parts, non-personal) data for scientific research purposes and with the aim of fostering innovation. Their provisions may facilitate the compliance with Art. 10 of the AI Act when developing AI systems, as more data are available, especially high quality data, and mechanisms are in place to foster the compliance with the provisions of the GDPR and the respect of fundamental rights and principles.

4.6.2.1 An example of national regulatory effort - Italy and the proposal law n. 1146 on AI

At the national level, most Member States are currently discussing national legislations for regulating the development and use of AI systems or appointing national authorities responsible for guiding stakeholders or enforcing existing norms.

In Italy, a Draft Law on AI (n. 1146) has been proposed, recently approved by the Senate of the Republic and (at the time of writing) it is being discussed for its final approval by the Chamber of Deputies in June 2025.

The Draft Law establishes various norms that aim at regulating both the development and use of AI systems, dedicating two articles, namely Art. 7 and Art. 8, to the specific context of healthcare. On the one hand, after having established that the use of artificial intelligence systems contributes to the improvement of the healthcare system, as well as to the prevention and treatment of diseases, while respecting individual rights, freedoms, and interests, Art. 7 codifies the right of patients to be informed about the use of artificial intelligence technologies during and for the provision of healthcare treatments, and that AI systems should always serve as a support in the decision-making process and should not affect the final decision of the healthcare professional, who ultimately remains the one responsible for the service.

On the other, focusing on the development and training of AI systems, Art. 8 partially derogates from the national data protection legal framework by establishing lighter rules for the processing of personal and non-personal data for research for the development of AI systems to be used in healthcare, when performed by specific actors, namely public and private non-profit entities, by the Scientific Institutes for Research, Hospitalization, and Healthcare (so called IRCCS under Italian law), as well as by private entities operating in the healthcare sector within research projects involving the mentioned entities. For the same purposes, the Italian legislator proposed to enable also the secondary use of personal and sensitive data, providing that the data subjects have been adequately informed, eventually by means of a general notice, and that the direct identifiers have been removed from these data. The conditions for availing oneself of this lighter regime include obtaining the approval of the Ethics Committee for the project and communicating such processing to the Italian Data Protection Authority. Further innovating on this point, the legislator proposed a mechanism of «silence means assent» if the Authority does not reply within 30 days.

The provisions outlined above constitute a significant innovation within the Italian legal framework. Historically, the Italian legislation concerning the processing of personal data, also the one that regulated the processing for scientific research purposes, has always adopted a rather conservative approach, emphasizing the importance of consent as the preferred legal basis and strictly limiting instances of further processing of personal data. The newly introduced provisions thus seem to mark a departure from this approach, establishing lighter provisions for the use of personal data, including health data, where the purposes pursued are deemed to be of substantial public interest. Among the latter, the legislator aims at expressly including artificial intelligence, recognizing its considerable potential to impact the provision of healthcare services.

The framework so envisioned, if definitely approved, aligns with the direction taken at the European level under the European Data Strategy, wherein AI-related purposes justify greater data accessibility, and its impact will be further strengthened and amplified by the concrete implementation of the latter, thus constituting a clear signal to stakeholders engaged in the development of artificial intelligence systems.

4.7 Liability in the context of AI

As mentioned in the previous pages, in cases where AI systems are implemented, and especially for the provision of healthcare services, addressing the question of allocating liability in case of damage caused to the patient or by the use of AI may be particularly challenging. More generally, the issue of allocating responsibilities for wrongdoings when AI is used is one to which is devoted particular attention.

To address the matter, the European Union first published the resolution Civil Law Rules on Robotics: European Parliament resolution of 16 February 2017 with recommendations to the Commission on Civil Law Rules on Robotics. Coherently with the pathway to adopting binding legal instruments in the field of new technologies (including AI), as described in the previous pages, the Resolution first of all highlighted the need to ascertain and evaluate whether the existing legal framework on liability was appropriate and sufficient «or whether [AI] calls for new principles and rules to provide clarity on the legal liability of various actors concerning responsibility for the acts and omissions of robots where the cause cannot be traced back to a specific human actor and whether the acts or omissions of robots which have caused harm could have been avoided» (Section AB).

The subsequent Report from the European Commission on the safety and liability implications of Artificial Intelligence, the Internet of Things, and robotics of 2020[99] precisely established that some of the core characteristics of AI, among which autonomy, data dependency, opacity, and overall complexity may be mentioned, may create new challenges in terms of safety and, in turn, of liability. As a consequence, and to address «potential uncertainties in the existing framework, [the European Commission called for] certain adjustments to the Product Liability Directive and national liability regimes through appropriate EU initiatives [...] a targeted, risk-based approach, i.e. taking into account that different AI applications pose different risks».[100]

As a consequence, the European co-legislators have worked in two different directions to address the issue. On the one hand, in September 2022, the European Commission published the Proposal for a Directive of the European Parliament and of the Council on adapting non-contractual civil liability rules to artificial intelligence (AI Liability Directive) and on the other recently updated its norms on liability for defective products (New Product Liability Directive) on November 18, 2024 with entry into force on December 8, 2024.

The Proposal originated from the realization and awareness that the liability rules in force in the Member States «are not suited to handling liability claims for damage caused by AI-enabled products and services.» Indeed, the various elements that, together or separately, contribute to the blackbox (ethical and legal) issue of AI, i.e., complexity, autonomy, and opacity, render it difficult and sometimes «prohibitively expensive for the victims» to prove the wrongfulness of an action or an omission of the system in court,[101]

---

[99] European Commission. Report from the Commission to the European Parliament, the Council and the European Economic and Social Committee. Report on the safety and liability implications of Artificial Intelligence, the Internet of Things and robotics, https://ec.europa.eu/info/sites/info/files/report-safety-liability-artificialintelligence-feb2020_en_1.pdf; 2020.

[100] *Ibid.*

[101] European Commission, *Explanatory Memorandum to Artificial Intelligence Act*, 2021.

and thus unsuited for the purpose of effectively claiming damages under the current applicable national liability rules. Indeed, from considering the difficulties at times encountered by the experts in the field in establishing precisely the causal link between input data and the output provided by the algorithm, it can be easily inferred that the burden of proving fault could easily turn into *probation diabolica*.

As a consequence, the objective of the Proposal was mainly twofold: «(a) the *disclosure of evidence* on high-risk artificial intelligence (AI) systems to enable a claimant to substantiate a non-contractual fault-based civil law claim for damages; (b) the *burden of proof* in the case of non-contractual fault-based civil law claims brought before national courts for damages caused by an AI system.»

The tools identified to reach the mentioned goals (a–b) were (a) the order for disclosure of evidence; (b) the presumption of existence of the causal link in case of fault based-liability (*presumption of causality*).[102]

As for the first one, the Commission proposed to provide national courts with the power to order to the provider of an AI system «to disclose [necessary and proportionate] evidence at its disposal about a specific high-risk AI system that is suspected of having caused a damage,» after having evaluated the plausibility of the claim and under the condition of having already autonomously asked for such information and having been refused.[103] Entitled to ask for such a measure were natural or legal persons with a direct sufficiently strong legitimate interest in having technical or otherwise (kept) confidential information disclosed, and therefore either a potential or actual claimant or a person subject to the obligation of a provider according to the AI Act or a user of the system. To sanction possible non-compliances with an order in this sense of a national court, the Proposal introduced a presumption of non-compliance with a relevant duty of care on the part of the provider and that the evidence requested was indeed proof of the damages potentially claimed.

On the other hand, the second tool consisted of a presumption of existence of the causal link between the fault of the provider and the output produced by the AI system (or the failure to produce any, as otherwise instructed). The Proposal set forth three conditions that should have been simultaneously met in order for this presumption to apply: (b1) the claimant had demonstrated the fault of the provider or the court had presumed it

---

[102] Barry Solaiman, «From 'AI to law' in healthcare: the proliferation of global guidelines in a void of legal uncertainty», *Medicine and Law*, 42(2), 2023, pp. 391-406.

[103] Article 3, paragraph 1.

after a refusal to comply with an order of disclosure; (b2) it was reasonably likely that such a fault had influenced the output produced by or the failure of the AI, concretely assessed on a case by case basis; (b3) the claimant had demonstrated the causal link between the output produced by or the failure of the AI and the damage caused.[104] The Proposal then went on to illustrate the possible cases that may involve fault on the part of the provider, basically related to violation of requirements set forth by the AI Act (transparency, quality of training dataset, failure of ensuring human oversight, no corrective actions taken when necessary).

Differently from the described Proposal, the recently published Directive 2024/2853 aims at updating from the previous Directive 85/374 the legal regime applicable to the issue of liability for defective products. As explicitly established in Article 1 the directive aims to «[lay] down common rules on the liability of economic operators for damage suffered by natural persons and caused by defective products, and on compensation for such damage.» To reach this end, and to adapt the previous norms to technological developments, the European co-legislators widened its material scope and thus its definition of «product» to newly include software and explicitly clarified that providers of artificial intelligence systems are to be considered manufacturers within the meaning of the New Product Liability Directive and are thus subject to its provisions. However, its norms will be applicable from December 9, 2026, the deadline for Member States to transpose them into national law, and only to products placed on the Union market from that moment onwards.

While both instruments clearly show the European Commission's attention to the issue of allocating responsibilities in cases of damage to patients, as mentioned, the legislative pathway for the AI Liability Directive was abandoned in early 2025. As a consequence, it will remain to be seen how the European co-legislators will address the mentioned issues and if a different specific set of norms will be enacted.

---

[104] Article 4, paragraph. 1.

4.8 AI Act and the Brussels effect?

As mentioned, the AI Act is frequently addressed as the first comprehensive regulation of artificial intelligence. When fully applicable, its norms will have an impact not only on the European market and the relevant stakeholders active in the European territory but also extraterritorially,[105] given that when defining the territorial scope of application of the AI Act establishes that it applies to «(a) providers placing on the market or putting into service AI systems or placing on the market general-purpose AI models in the Union irrespective of whether those providers are established or located within the Union or in a third country» and «(c) providers and deployers of AI systems that have their place of establishment or are located in a third country, where the output produced by the AI system is used in the Union.» Indeed, the only way to fully reach the aim of the AI Act is to enact norms that are applicable to any AI system that may potentially endanger people in the European Union and/or European values, irrespectively of the place of establishment or residence of the natural or legal person that provides the AI system itself. Therefore, the focus is twofold, i.e., on the place of putting into service or of placing on the market of the AI system or on the output it produces.

This element of the AI Act architecture, along with its primacy, frequently raises the question of the potential for a *Brussels effect* in this field, similar to what happened for the data protection regulation and approach.[106] The term was first coined by Anu Bradford in 2012 to highlight and describe the extraterritorial power of EU norms in the fields of data protection, environmental law, and antitrust.[107] More specifically, it is the capacity of EU

---

[105] Graham Greenleaf, «EU AI Act: Brussels effect(s) or a race to the bottom», *190 Privacy Laws & Business International Report* 1, 2024, pp. 3-6.

[106] Carl Siegmann, and Martin Anderljung, «The Brussels effect and artificial intelligence», 2022, https://www.governance.ai/research-paper/brussels-effect-ai; Fabiana Lütz , «How the 'Brussels effect' could shape the future regulation of algorithmic discrimination», *Duodecim Astra*, 1, 2024, pp. 142-163; Gerhard Wagner, «Liability rules for the digital age: aiming for the Brussels effect», *Journal of European Tort Law*, 13(3), 2023, pp. 191-243. Of the opposite view, Alice Engler, «The EU AI Act will have global impact, but a limited Brussels effect», 2022, available at https://www.brookings.edu/research/the-eu-ai-act-will-have-global; Ugo Pagallo, *Why the AI Act won't trigger a Brussels effect*, cit.; Marco Almada, Anca Radu, «The Brussels side-effect: how the AI Act can reduce the global reach of EU Policy», *German Law Journal*, 25, 2023, pp. 646-663; Michal Czerniawski, «Towards the effective extraterritorial enforcement of the AI Act», in Jaap-Henk Hoepman *et al.*, Meiko Jensen, Maria Grazia Porcedda, Stefan Schiffner, Sébastien Ziegler (ed. by), *Privacy symposium 2024—Data protection law international convergence and compliance with innovative technologies (DPLICIT)*, Springer, 2025, pp. 35-53.

[107] Anu Bradford, «The Brussels effect», *Northwestern University Law Review*, 107(1), 2012, pp. 1-68; Anu Bradford, *The Brussels effect: How the European Union*

regulations, especially in the mentioned fields, to trigger a «phenomenon of globalised regulation (...) through legal means, political influence or market mechanisms».[108] After the publication of the proposal for a comprehensive regulation on AI, questions were raised regarding the possibilities for AI to be included among the fields for which EU norms may have extraterritorial effects of the mentioned kind.

Indeed, the AI Act may have both *de facto* and *de jure* effects on other jurisdictions, which are usually identified as the typical elements through which the effect occurs. As for the first one, and concentrating on potential market mechanisms, it is postulated that non-EU companies that conduct their business both on the European market and outside of it may decide to design and develop a single version of their AI system compliant with the (already enacted and applicable) EU norms and then offer it also on other markets worldwide, in order not to have two different systems of their product in place and because of the «high differentiation costs».[109] Indeed, this approach was adopted by Facebook, Sonos, and Microsoft for their customers worldwide when they decided to implement GDPR-compliant data protection systems.[110] Moreover, the EU has «favourable market properties»,[111] such as the establishment of regulatory sandboxes that may attract companies from other jurisdictions and incentivize the development and marketing of AI Act-compliant AI systems. At the same, it may be expected and anticipated that companies worldwide would have to justify the decision not to adopt a standard not as high as the European one.[112] On the other hand, the *de jure* effect of the AI Act would entail the adoption of EU-inspired regulations of AI systems around the globe. Indeed, it has been highlighted that the proposed AI regulation in Brazil already envisions a risk classification similar to the one adopted in the AI Act,[113] and that China may decide to adopt a legislation inspired by the European one.[114] While there is still little

rules the world, Oxford University Press, 2020.

[108] Fabian Lütz, «How the 'Brussels effect' could shape the future regulation of algorithmic discrimination», *Duodecim Astra*, 1, 2021, pp. 142-163.

[109] Juliette Faivre, «The AI Act: towards global effects?», *European Science, Technology and Innovation Policy*, 2023, pp. 1-11.

[110] Graham Greenleaf, *op. cit.*

[111] Carl Siegmann, and Martin Anderljung, «The Brussels effect and artificial intelligence», *Centre for the Government of AI*, 2022, available at https://www.governance.ai/research-paper/brussels-effect-ai

[112] Luciano Floridi, *op. cit.*

[113] Benjamin Kohn, «AI regulation around the world», 2023, available at https://www.taylorwessing.com/en/interface/2023/ai---are-we-getting-the-balance-between-regulation-and-innovation-right/ai-regulation-around-the-world; Juliette Faivre, *op. cit.*

[114] Luciano Floridi, *op. cit.*

evidence of the possible concrete influence of the EU provisions on other regulations on the matter under development worldwide, in order for the Brussels effect to be effective the core of the regulation should be replicated in other jurisdictions, such as in particular the risk-based approach and the related classifications of the AI systems.[115]

On the contrary, authoritative voices that claim that a Brussels effect is «unlikely» to happen base their assumption on exogenous and endogenous reasons.[116] As for the first, it should be kept in mind that the European Union is not the only jurisdiction aiming at enacting norms and a comprehensive regulation on AI, as previously already extensively described and further analyzed in the following pages. Each of these legal systems presents its own peculiarities, traditions and dissimilarities, which may «exacerbate current disparities of legal approaches to AI».[117] As a consequence of these «crucial differences,» as well as the fact that the European approach involves «an odd mix of hard law and soft law, co-regulation and self-regulation, horizontal and vertical provisions,» it is much more likely that, if any, a new Brussels effect would involve at most specific provisions of the AI Act and not its core pillars.[118] Concerning the endogenous reasons, it remains to be seen if the definitions and provisions of the AI Act will pass the *test of time*, and thus if they are future-proof enough to adapt to the technological development in the field, as well as adequately conceptualized.[119]

Moreover, it should also be kept in mind that the AI Act and the GDPR have not been conceptualized in the same manner by the European co-legislators. Indeed, differently from the latter, the AI Act, as mentioned, is «part of a more extensive legislative ecosystem»,[120] which comprises various normative acts that complement and implement the provisions recently specifically dedicated to AI systems, and as a consequence «transplanting» the AI Act alone might not be feasible.

More generally, while on the one hand, the advantages of a Brussels effect for the AI Act inevitably include the improvement of the level of harmonization of the regulation on the matter worldwide, possible disadvantages are related to the fact that for some jurisdictions, the European Union usually

---

[115] Graham Greenleaf, *op. cit.*

[116] Ugo Pagallo, *Why the AI Act won't trigger a Brussels effect*, cit.

[117] *Ibid.*

[118] *Ibid.*

[119] *Ibid.*

[120] Michal Czerniawski, «Towards the effective extraterritorial enforcement of the AI Act», in Jaap-Henk Hoepman *et al.* (ed. by), *Privacy symposium 2024—Data protection law international convergence and compliance with innovative technologies (DPLICIT)*, Springer, 2025 forthcoming.

adopts an over-regulatory regime, which may be seen as problematic for the development of the technology and the market.[121] Therefore, it remains to be seen how much the AI Act will influence future regulation of AI world-wide. For now, in the following pages, we will provide an overview of the current approaches under development worldwide.

---

[121] Juliette Faivre, *op. ct.*; Alex Engler, «The EU AI Act will have global impact, but a limited Brussels effect», 2022, available at https://www.brookings.edu/articles/the-eu-ai-act-will-have-global-impact-but-a-limited-brussels-effect/.

5 Maintaining a Global Leadership: Regulation of Artificial Intelligence in the United States

5.1 Betting on artificial intelligence to develop cutting edge innovation in healthcare and maintain US leadership in the global economy

The United States, home to many top AI companies, is pivotal in setting standards and developing strategies to overcome challenges and leverage the substantial opportunities presented by artificial intelligence. The US strategy has been cautious, aiming to avoid burdening companies in a manner that might hinder innovation, as the key goal for the country is to uphold its dominance in the AI sector. To succeed in this, especially in the face of worldwide competition, the United States significantly depends on private-sector innovation, complemented by public investment to support these efforts and nurture a vibrant AI ecosystem.

The use of artificial intelligence in the healthcare sector holds significant promise, particularly in the United States, where the healthcare system is facing significant challenges. In 2022, healthcare expenditures in the US climbed to \$4.5 trillion, averaging \$13,493 per individual.[1] This figure notably exceeds the per capita healthcare costs in other prosperous nations by more than a factor of two.[2] The US healthcare system is also facing various challenges, such as an elderly demographic, escalating comorbidity rates, the aftermath of the pandemic, and the ever-increasing expenses associated with healthcare delivery.[3] However, there's optimism that AI can enhance the efficacy of the healthcare system, whether it's through clinical AI applications assisting in projecting patient health outcomes and suggesting treatments, or administrative AI solutions aimed at lightening the workload on healthcare providers and streamlining operational workflows.[4]

---

[1] Micah Hartman, *et al.*, «National healthcare spending in 2022: growth similar to prepandemic rates», *Health Affairs,* 43(1), Pharmaceuticals, Opioid Use, Health Spending & More, 2021.

[2] Peter G. Peterson Foundation, *Why Are Americans Paying More for Healthcare?*, Foundation website, 2024.

[3] Charles H. Jones, and Mikael Dolsten, «Healthcare on the brink: navigating the challenges of an aging society in the United States», *NPJ Aging,* 10(1), 22, 2024.

[4] Adam Bohr, and Kaveh Memarzadeh, «The rise of artificial intelligence in healthcare applications», *Artificial Intelligence in Healthcare*, 2020, pp. 25-60.

Bearing this in mind, it also becomes clear why corporate entities can so effectively influence conversation. Their substantial prominence and resources are among the nation's most valuable assets, with projections indicating that by 2030, AI could contribute to a remarkable 21% net increase in the US GDP, thus highlighting its role in economic advancement.[5] Thus, the crucial challenge for the nation is to implement adaptable regulations tailored to specific applications, which can be quickly adjusted to future developments. Consequently, also US policymakers have favored a risk-based approach, but, in contrast to the European Union, steering away from broad AI regulation in favor of industry-specific guidelines, entrusting responsibility to federal agencies.

5.2 Key differences between the European Union and the United States AI regulatory frameworks

The United States plays a dual role in comparison to Europe when it comes to artificial intelligence. On one hand, the US dominates in the AI sector, thanks to significant investments and developments, while they are home to major AI companies. On the other hand, the absence of federal legislation and a comprehensive policy framework results in the various states often pursuing different regulatory tactics for particular industries or uses of AI, creating an inconsistent legal landscape. Thus, the challenge for the US lies in harmonizing the disparate state regulations into a unified federal strategy that also takes the AI industry's interests into account is the US's difficulty. Finding this balance is not straightforward and may be a drawn-out process given the influence of the industry within the American political system.

Initially, the US and EU strategies appear to have many commonalities, such as endorsing and adopting a risk-based approach, agreeing on a comparable definition of trustworthy AI, and supporting international cooperation (notably their involvement in the G7 and the Trade and Technology Council).[6] But upon further examination, the two frameworks for AI risk management reveal numerous differences.

In fact, the strategies of the United States and the European Union distinctly reflect their varied objectives: the US is deeply invested in AI research, development, and implementation to retain its global dominance,

---

[5] Katherine Haan, and Rob Watts, *How Businesses Are Using Artificial Intelligence In 2023*, Forbes Advisor, 2023.

[6] For example, the G7 Leaders' Statement on the Hiroshima AI Process and the agreement reach at the Fourth Ministerial meeting of the Trade and Technology Council (TTC), October 30, 2023.

while the EU is concerned with safeguarding transparency and the privacy of its citizens. Accordingly, America has opted for a sector-specific regulatory methodology, whereas Europe has implemented a unified horizontal framework that brings uniformity to AI regulations across its member states. While, pursuant to Executive Order 14110, the US prioritized establishing AI standards and practices mainly for key federal departments, the European AI Act imposes stringent, obligatory guidelines, positioning itself as one of the most rigorous AI regulatory systems worldwide. On one hand, even though US federal agencies are progressively updating their policies to accommodate AI advancements, they may not possess the authoritative power to enforce these rules as solidly as Europe, which possesses a stronger centralized mandate and additional funding, laying the groundwork for a robust enforcement mechanism.[7] On the other side, the European method promotes consistency but may lack adaptability and could face unanticipated regulatory voids as technology advances.

5.3 The long process to build a federal regulatory framework

Starting with President Obama, all most recent presidents of the United States were confronted with the urge to design a federal framework for AI regulation and provide guidance to the nation. While President Obama opted for a principle-based approach, balancing innovation with public safety, discussing specifically the risks of bias and discrimination and including multiple stakeholders in its development, President Trump favored a more market-driven and reactive approach, focusing on national security, global competition (particularly in relation to China), and boosting AI use in the public and private sector. Last, President Biden came back to President Obama's approach, building on his previous work emphasizing proactive governance and ethical considerations, while providing a more comprehensive framework for AI through Executive Orders.

5.3.1 President Obama administrations

During his presidency (2009–2017), particularly in its latter part, President Obama identified the potential of AI and focused on its societal implications, initiating efforts to engage with industry leaders, researchers, and civil society

---

[7] Alex Engler, *The E.U. and U.S. diverge on AI regulation: A transatlantic comparison and steps to alignment*, Brookings, April 25, 2023.

to assess risks related to AI, including bias and discrimination. His administration aimed to establish a framework for AI governance that balanced innovation with public safety, providing the principles on which a more complex governance could have been developed. By fostering a collaborative environment, multiple perspectives were considered in policy decisions aimed at addressing concerns about bias, discrimination, and safety. President Obama's focus was on developing responsible AI policies with attention to ethical implications and societal impacts, aiming to ensure that advancements in AI benefit all communities, especially underserved populations. These initiatives set the stage for future policies prioritizing health equity in AI applications.

The Obama administration took early steps to orchestrate a federal approach to the emerging influence of AI by having the National Science and Technology Council release a report in October 2016 titled «Preparing for the Future of Artificial Intelligence.»[8] This report explored the state of AI, its practical applications, and the societal challenges it poses, including justice, fairness, accountability, and safety. It also considered how AI's growth would affect policies in areas like international relationships, cybersecurity, and defense. The report also recommended actions for Federal agencies and other stakeholders, and it emphasized the role of AI in serving the public good, noting that investments in basic and applied research in AI have led to notable advancements, especially in healthcare. An example provided was Walter Reed Medical Center's implementation of AI by the Department of Veteran Affairs to foresee medical complications and improve the treatment of severe combat injuries, which resulted in better patient outcomes, accelerated healing, and lowered costs. Additionally, the report highlighted efforts at Johns Hopkins University, where predicting problems to provide preemptive care reduced hospital-acquired infection rates. Finally, the document concluded that, as health records become digitized, the predictive analysis of health data could play an essential role in various healthcare sectors, including precision medicine and cancer research. The report built on three previous White House reports, from 2014 (which devoted some pages to the use of open data in healthcare, the «blue button» initiative, the use of big data in healthcare delivery, and called for harmonization of privacy frameworks),[9] 2015,[10] and 2016,[11] on big data and algorithmic systems.

---

[8] Executive Office of the President National Science and Technology Council, Committee on Technology, *Preparing for the future of artificial intelligence*, 2016.

[9] Executive Office of the President, *Big data: Seizing opportunities, Preserving values*, 2014.

[10] Executive Office of the President, *Big data: Seizing opportunities, Preserving values*, 2015.

[11] Executive Office of the President, *Big Data: A Report on Algorithmic Systems,*

A day following the associated report, the «National Artificial Intelligence Research and Development Strategic Plan»[12] was unveiled. This strategy outlined seven key R&D efforts to be pursued, including making long-term investments in AI research; developing effective methods for human-AI collaboration; understanding and addressing the ethical, legal, and societal implications of AI; ensuring the safety and security of AI systems; developing shared public datasets and environments for AI training and testing; measuring and evaluating AI technologies through standards and benchmarks; better understanding the national AI R&D workforce needs. In the healthcare field, the plan recognized AI's potential to aid bioinformatics systems in identifying genetic risks from vast genomic studies and in predicting the safety and efficacy of novel pharmaceuticals.

The Plan was first updated in 2019[13], in which occasion a new strategy, focused on expand public–private partnerships to accelerate advances in AI, was added, and in 2023,[14] which saw the inclusion in the Plan of the ninth strategy, establishing a principled and coordinated approach to international collaboration in AI research.

5.3.2 President Trump I administration

The Trump administration, from 2017 to 2021, focused on AI's national security implications, particularly in competition with countries like China. His approach was reactive, addressing more immediate threats than focusing on developing long-term regulatory frameworks. Unlike Obama, the Trump administration engaged fewer stakeholders and emphasized US competitiveness in AI through a market-driven approach. Trump viewed AI mainly from a national security and global competition perspective, issuing executive orders to promote innovation and undertaking initiatives to reinforce US leadership in the AI space. As a result, many of President Trump's proposals did not include thorough guidelines for ethical use and societal impacts, diverging from the approach taken by the previous administration.

Opportunity, and Civil Rights, 2016.

[12] National Science and Technology Council and the Networking and Information Technology Research and Development Subcommittee, *National Artificial Intelligence Research and Development Strategic Plan*, 2016.

[13] Select Committee on Artificial Intelligence of The National Science and Technology Council, *National Artificial Intelligence Research and Development Strategic Plan*, 2019.

[14] Select Committee on Artificial Intelligence of The National Science and Technology Council, *National Artificial Intelligence Research and Development Strategic Plan*, 2023.

In February 2019, President Trump enacted Executive Order (EO) 13859, titled «Maintaining American Leadership in Artificial Intelligence.» This order was geared towards creating foundational guidelines and strategies to advance the United States' proficiency in artificial intelligence with objectives of enhancing scientific research, competitive strength in the economy, and national security. The order initiated the American AI Initiative, a comprehensive strategy devised to elevate the United States to the forefront of AI leadership and covered five critical areas: investing in AI Research and Development; unleashing AI resources; setting AI governance standards; building the AI workforce; international engagement and protecting American AI Advantage.

Among other requirements, the Executive Order directed the Director of the Office of Management and Budget (OMB) to issue guidance for all federal agencies to shape regulatory and nonregulatory strategies concerning technology and industry sectors boosted by artificial intelligence, also considering ways to minimize impediments to AI technology's development and uptake. The issued memorandum[15] in November 2020 mandated that when creating regulations or policies pertinent to AI applications, agencies should also support advancements in technology and innovation. It was thus imperative that they protected American technological, economic, and national security, along with privacy, civil liberties, and values such as freedom, human rights, rule of law, and intellectual property protection. Furthermore, the memorandum clarified that determining the right regulatory or nonregulatory approach to privacy and other risks fundamentally relies on the risk nature and available mitigation tools, hence introducing a risk-focused framework without instituting strict mandates.

Also, during his administration, President Trump committed to doubling AI research investment[16] and establishing the first-ever national AI research institutes,[17] issued a plan for AI technical standards,[18] and established guidance for Federal use of AI, with the Executive Order on Promoting the Use of Trustworthy Artificial Intelligence in the Federal Government. The

---

[15] Executive Office of the President, Office of Management and Budget, *Guidance for Regulation of Artificial Intelligence Applications, Memorandum for the Heads of Executive Departments and Agencies*, 2020.

[16] *President Trump's FY 2021 Budget Commits to Double Investments in Key Industries of the Future*, Trump White House Archives, 2020.

[17] *The Trump Administration Is Investing $1 Billion in Research Institutes to Advance Industries of the Future*, Trump White House Archives, 2020.

[18] National Institute of Standards and Technology, *U.S. leadership in AI: A Plan for Federal Engagement in Developing Technical Standards and Related Tools*, submitted on 9th August 2019.

Executive Order laid out guidelines for AI usage within the government, introducing a unified approach to embrace these principles, commanded governmental bodies to inventory their AI applications, and tasked the General Service Administration alongside the Office of Personnel Management with boosting AI integration competencies across agencies. In line with the National AI Initiative Act of 2020, and responding to this strategic direction, the White House early in January 2021 founded the National Artificial Intelligence Initiative Office within the Office of Science and Technology Policy (OSTP). This office continues to act as the nerve center for the coordination and cooperative efforts in AI-related research and policy creation throughout the US government, combining efforts with the private sector, academic institutions, and other relevant parties.

Finally, in September 2019, the White House convened the Summit on Artificial Intelligence in Government to collect insights on how federal agencies could integrate AI to fulfill their missions and enhance public services. To fast-track AI implementation within government bodies, the General Services Administration initiated the AI Center of Excellence in September 2019, followed by the formation of the AI Community of Practice. Subsequently, the AI Center of Excellence was enshrined in legislation through the AI in Government Act of 2020.

5.3.3 President Biden administration

From 2021 to 2024, President Biden administration expanded on Obama's legacy, addressing similar concerns and even seeking input directly from him, for example when drafting executive orders.[19] For instance, the principles from Obama's 2016 AI report, which emphasized proactive governance and ethical considerations, continued to influence Biden's policies; both administrations stayed committed to tackling AI risks, such as bias and discrimination, and have sought input from various stakeholders, ensuring diverse perspectives in shaping AI regulations. This approach was aimed at building public trust in AI systems while maintaining US leadership in global AI policy. Overall, AI policy and regulation saw significant developments under President Biden.

In October 2022, the White House issued «The Blueprint for an AI Bill of Rights,»[20] a set of five principles (safe and effective systems; algorithmic

---

[19] Liz Landers, and Luke Barr, *How Obama helped President Biden draft the AI executive order*, ABC News, November 3, 2023.
[20] White House Office of Science and Technology Policy, *Blueprint for an AI Bill of Rights, Making Automated Systems Work for The American People*, October 2022.

discrimination protections; data privacy; notice and explanation; human alternatives, consideration, and fallback) and associated practices to help guide the design, use, and deployment of automated systems to protect the rights of the American public in the age of artificial intelligence. The AI Bill explicitly mentioned healthcare as an opportunity for AI applications,[21] while listing some potential risks of AI misuse in the sector. For instance, when tackling biases in algorithms, one example cited was a healthcare predictive tool that consistently gave Black patients lower risk scores compared to white patients, even though both groups had similar chronic conditions and equivalent health indicators.[22] Additionally, in the context of discussing principles of «human alternatives, consideration, and fallback,» it was noted that during the 2022 plan year, the administration under President Biden and Vice President Harris boosted investments, enabling grantee organizations to train and certify over 1,500 navigators (professionals who assist consumers, small companies, and their staff in exploring their health insurance options) to aid uninsured individuals in securing affordable yet comprehensive health insurance coverage.

In February 2023, revisiting the matters of prejudice and inequality, President Biden initiated the Executive Order on Advancing Racial Equity and Assistance for Underserved Communities via the Federal Government, which «directs federal agencies to root out bias in their design and use of new technologies, including AI, and to protect the public from algorithmic discrimination.»

To foster reliable AI usage by private companies, the Biden administration obtained pledges in July 2023 from seven leading Artificial Intelligence firms (Amazon, Anthropic, Google, Inflection, Meta, Microsoft, and OpenAI) to endorse the safe and ethical advancement of AI.[23] These pledges involved conducting security evaluations, exchanging risk management data, investing in cybersecurity, reporting on AI capabilities, and researching how to reduce potential societal harm. Subsequently, other enterprises such as Adobe, Cohere, IBM, Nvidia, Palantir, Salesforce, Scale AI, and Stability also voluntarily committed to the White House to advance the secure, transparent, and responsible development and application of AI. Nonetheless, the importance of involving US corporations in policy-making is undeniable to ensure cooperation and effectiveness of

---

[21] *Ibid.*, p. 10.

[22] *Ibid.*, p. 25.

[23] White House, *Fact Sheet: Biden-Harris Administration Secures Voluntary Commitments from Leading Artificial Intelligence Companies to Manage the Risks Posed by AI*, Statements and releases, July 21, 2023.

reforms, with a continuous exchange between businesses and governmental bodies being essential. For example, in February 2024, the Biden administration launched a nation-wide effort promoting the secure advancement and application of generative AI through the new US AI Safety Institute Consortium (AISIC), operating under the auspices of the US AI Safety Institute (USAISI). Over 200 companies have joined the consortium, including prominent AI firms such as OpenAI, Alphabet, Anthropic, Microsoft, Meta Platforms, Apple, Amazon, and Nvidia. Additionally, other entities touched by the waves of AI innovation such as Palantir, Intel, JPMorgan Chase, Bank of America, Cisco Systems, IBM, Mastercard, and Visa have become part of this initiative.

A significant achievement for this Administration was however set on October 30, 2023, when President Joe Biden issued Executive Order (EO) 14110 titled «Safe, Secure, and Trustworthy Development and Use of Artificial Intelligence,»[24] urging for a proactive approach to managing the risks associated with AI and machine learning technologies as they were becoming more integrated into the healthcare sector. The order aimed at leveraging AI positively to maximize its numerous advantages while also curbing significant risks, highlighting eight key policy areas that should guide the US institutions broadly (in anticipation of further Congressional action to provide more comprehensive and authoritative regulations): safety and security; innovation and competition; worker support; addressing AI biases and civil rights concerns; consumer protection; privacy; federal deployment of AI; and asserting international leadership. Moreover, it instructed federal agencies to increase supervision of AI applications within their domains while the White House collaborates with Congress to seek bipartisan legislative solutions. The executive order delegated responsibility to over 50 federal agencies, assigning them more than 100 specific directives typically meant for execution within less than one year's time. It also established a White House AI Council, consisting of leaders from 28 federal departments and agencies to ensure organized enforcement. Thus, the executive order provided a structured agenda for forming a regulatory and administrative framework that promotes and controls the progression of artificial intelligence efficiently.

---

[24] The definition of AI used in this EO is derived from the National AI Initiative Act of 2020—«a machine-based system that can, for a given set of human-defined objectives, make predictions, recommendations, or decisions influencing real or virtual environments. Artificial intelligence systems use machine- and human-based inputs to perceive real and virtual environments; abstract such perceptions into models through analysis in an automated manner; and use model inference to formulate options for information or action» 15 USC 9401(3).

The primary objective of the executive order was to balance risk management with the promotion of innovative AI applications that could benefit consumers. To this end, the executive order issued healthcare-specific mandates for the US Department of Health and Human Services (HHS) aimed at ensuring the secure and ethical application of AI technologies in the medical sector. These included the creation of an AI Task Force tasked with developing policy to foster regulatory clarity and spur AI advancements in healthcare systems.[25] The Task Force began coordinating efforts to formulate principles for combating racial bias in healthcare algorithms. The directives also called on the Department of Health and Human Services to: identify and give precedence to funding awards (and related initiatives) for the responsible progression of AI technology; devise a strategy addressing state and local use of automated or algorithmic systems in the allocation of public benefits and services funded by the department; create a plan to evaluate AI quality, encompassing the establishment of an AI assurance policy and determining infrastructure requirements; review potential measures to increase industry understanding of and adherence to federal anti-discrimination laws applicable to AI within health and human service sectors receiving government funds; initiate an AI safety program which will oversee and refine AI implementation in healthcare via a unified framework for pinpointing AI-related clinical errors and protocols for averting such issues; outline a strategy for employing AI and AI-empowered tools in drug discovery processes; propose a system for providers of synthetic nucleic acids to advocate the incorporation of procurement screening processes, including industry standards and incentives; and finally, generate specifications, guides, and best practice recommendations for sequence-of-concern database management along with conformity assessments, all for the utilization by providers of nucleic acid sequences.

In October 2024, building on the Executive Order, President Biden released the first-ever National Security Memorandum (NSM) on Artificial Intelligence,[26] to direct the US Government to implement concrete and impactful steps to (1) ensure that the United States leads the world's development of safe, secure, and trustworthy AI; (2) harness cutting-edge AI technologies to advance the US Government's national security mission; and (3)

---

[25] Marshall H. Chin, *et al.*, «Guiding Principles to Address the Impact of Algorithm Bias on Racial and Ethnic Disparities in Health and healthcare», *JAMA Netw Open*, 2023.
[26] White House, *Memorandum on Advancing the United States' Leadership in Artificial Intelligence; Harnessing Artificial Intelligence to Fulfill National Security Objectives; and Fostering the Safety, Security, and Trustworthiness of Artificial Intelligence*, 2024.

advance international consensus and governance around AI. Furthermore, a framework providing further detail and guidance to implement the Memorandum, including requiring mechanisms for risk management, evaluations, accountability, and transparency, was published alongside.[27] In one way, this Framework reflected the same approach of other international regulations, such as the European Union's AI Act, which took effect in August 2024. In fact, it outlined a risk-based approach to managing AI activities, with a specific focus on identifying and addressing high-impact AI use cases. Specifically, the Memorandum designated certain AI activities as high-impact, particularly those where AI's results serve as a core basis for decisions or actions that could threaten national security, international norms, human rights, civil liberties, privacy, safety, or other democratic values.[28] Also, the Framework mandated that federal agencies provide clear guidance on managing these high-impact use cases, ensuring that risk management practices are in place to mitigate potential harm. Furthermore, agencies are required to integrate privacy, civil liberties, and safety experts into their AI governance and oversight processes. This cross-disciplinary involvement helps ensure that AI systems are developed and deployed with appropriate safeguards to mitigate risks related to privacy violations, biases, and unintended consequences that could undermine public trust or compromise fundamental freedoms.

5.3.4 Key differences between the United States executive order on the safe, secure, and trustworthy development and use of artificial intelligence and the European Artificial Intelligence Act

While both the US and the EU opted to regulate AI within a similar period (2023 for Biden's Executive Order and 2024 for the adoption of the European AI Act), they have taken distinctly different approaches, beginning with their choice of regulatory instruments.[29] In fact, the AI Act is a regulation, meaning a binding legal framework developed by all the European co-legislators: it was proposed by the European Commission in April 2021, discussed by the European Parliament and the Council and then formal-

---

[27] White House, *Framework to Advance AI Governance and Risk Management in National Security*, 2024.

[28] The Memorandum explicitly states that «Such high-impact activities shall include AI whose output serves as a principal basis for a decision or action that could exacerbate or create significant risks to national security, international norms, human rights, civil rights, civil liberties, privacy, safety, or other democratic values».

[29] Camilla Scarpellino, «E.U. and U.S. regulatory approach to AI: a comparative perspective», *Luiss Policy Observatory*, 2024.

ly adopted by the European Parliament. It directly applies to all Member States, and it becomes legally binding two years after its publication in the Official Journal of the European Union. Conversely, the Executive Order in the United States is a directive issued by the President to influence the policy framework, which holds the force of law and does not require congressional approval.

The AI Act and the Executive Order differ also in their aims and scope. On one side, the AI Act seeks to harmonize regulations for AI systems in the European market, ensuring they comply with necessary requirements for legal use and trade within the EU On the other hand, the Executive Order aims to enhance competitiveness and leadership in technology by promoting responsible innovation that benefits public welfare; the Order focuses mainly on developing and deploying AI technologies across multiple sectors, highlighting the role of federal agencies in leading efforts and protecting rights without setting specific risk levels or restrictions associated with AI.

In the European Union, the oversight framework is centralized, beginning with providers' Risk/Impact Assessments and private certification. This process then escalates to the National Authority, which oversees and coordinates the certification bodies responsible for issuing these certificates. Consequently, each Member State must designate at least one notifying authority and one market surveillance authority. The notifying authority manages the bodies that assess AI devices' compliance with the AI Act and issues the conformity certification required by AI operators to market their devices and software in Europe. Conversely, in the United States, oversight through the Executive Order (EO) involves specific agencies tasked with studying and reporting on AI matters in critical infrastructure, cybersecurity, and chemical, biological, radiological, and nuclear (CBRN) threats. An AI Council, established by the White House, supervises the activities of these agencies.

Additionally, the EU's system of AI governance revolves around a centralized approach that integrates European institutions, national authorities, and certifying bodies, stipulating timely cooperation among all AI stakeholders, including manufacturers, deployers, importers, and users of AI products. Moreover, the AI Act creates the AI Office, a dedicated body for monitoring the implementation of the AI Act throughout the EU On the other hand, the executive order's governance framework in the US emphasizes the assessment of AI risks and opportunities within each Administrative Body's area of expertise.

Both the EU's AI Act and the US Executive Order address system testing and monitoring throughout an AI system's life cycle. The AI Act demanded that companies validate their compliance through extensive pre-market testing protocols (outlining the methods and metrics applied in pre-launch tests) and a post-market monitoring strategy (centered on the developer's oversight of ongoing system performance). The echoes this by noting that «testing and evaluations, including post-deployment performance monitoring, will help ensure that AI systems function as intended, are resilient against misuse or dangerous modifications, are ethically developed and operated in a secure manner, and are compliant with applicable Federal laws and policies.» Moreover, it particularly points to healthcare, stating «infrastructure needs for enabling pre-market assessment and post-market oversight of AI-enabled healthcare-technology algorithmic system performance against real-world data.»

5.3.5 The efforts of the US Congress in developing a comprehensive regulatory framework on AI and data protection

After providing directions to the agencies, President Biden's Executive Order on the Safe, Secure, and Trustworthy Development and Use of Artificial Intelligence specifically urged US Congress to enact bipartisan data privacy legislation. This underscored, once again, the importance of collaboration between the Government and Congress, as well as of the interplay between privacy laws and AI regulation is. However, the United States Congress has been notably proactive in attempting to regulate artificial intelligence, even if such a framework has not yet been approved; for instance, as of September 2024, more than 120 bills pertaining to the regulation of artificial intelligence were under consideration.

There are several reasons why Congress legislation would be preferable to a presidential executive order; in general, while an executive order like President Biden's one is able to set important guidelines for the responsible development of AI technologies, only congressional legislation is likely to provide a more effective and enduring framework for regulating AI. The fact that President Trump's first action in his second mandate was to issue a regulatory freeze that also impacted President Biden's Executive Order on the Safe, Secure, and Trustworthy Development and Use of Artificial Intelligence, which was rescinded, is a clear example of this situation. In fact, laws passed by Congress have a more permanent status than executive orders, which can be modified or rescinded by subsequent administrations (and they often are). They can also result from bipartisan proposals, leveraging broader support to create legislation that reflects a range of interests and concerns. This permanence is crucial for establishing long-term regulatory frameworks that guide AI development and deployment, ensuring policies remain in place regardless of political changes. Also, industries benefit from consistent guidelines that foster investment and are likely to prefer regulatory stability. Additionally, Congress has the capacity to create comprehensive legislation addressing various aspects of AI, including safety, privacy, consumer protection, and ethical considerations; this approach can be more effective than executive orders, which may focus on specific issues without addressing broader implications. Furthermore, legislative processes usually involve hearings and discussions that allow for input from multiple stakeholders, including industry experts, civil society, and affected communities, leading to balanced and effective regulations. Moreover, legislation typically includes mechanisms for enforcement and accountability, ensuring compliance among AI developers and users, while executive orders may lack these provisions, limiting their effectiveness. Lastly, congressional action can set federal standards, avoiding inconsistent state regulations, which is crucial to facilitating industries and businesses that may operate in more than one state.

Congres's attempts to legislate AI started several years ago. For example, during the 115th Congress spanning 2017–2019, Section 238 of the John S. McCain National Defense Authorization Act for Fiscal Year 2019 was passed, which mandated that the Department of Defense engage in a range of AI-centered initiatives, including designating a coordinator for such efforts. The momentum continued with the introduction of the AI Training Act by Senators Peters and Portman in 2021, a bipartisan effort that was passed into law in 2022 and which obligates the Office of Management and Budget to create or provide an AI education program for executive agencies' acquisition teams, with some exceptions.

After President Biden's Executive Order, there has been a consistent push for bipartisan legislation. A notable attempt was made by Senators Blumenthal and Hawley, with their No Section 230 Immunity for AI Act, proposed in June 2023. The aim of the proposal was to carve out an exception from Section 230 (which shields internet firms from being sued for user-generated content) specifically for artificial intelligence, by introducing a clause that would remove protections for AI companies in legal actions concerning the use of generative AI. This change would have enabled Americans affected by generative AI models to pursue legal action against AI companies in federal or state courts. Furthermore, in September 2023, the senators unveiled a bipartisan legislative blueprint designed to regulate artificial intelligence. The blueprint outlined several key principles that should have guided future laws, such as establishing an independent regulatory agency, ensuring liability for damages, safeguarding national security interests, enhancing transparency, and protecting consumers and children.

In June 2023, Congressman Lieu, Congressman Buck, and Congress-woman Eshoo presented another bipartisan proposal with the National AI Commission Act. This legislation, supported by both parties and across both chambers, aimed to establish a national commission dedicated to ad-dressing how Artificial Intelligence should be regulated. The Commission's role would have involved reviewing the United States' current strategy for AI oversight and providing suggestions on how to enhance and fortify those regulations. It would have also contemplated whether there are needs to in-stitute new governmental bodies or offices for the efficient administration of AI, alongside creating a risk-based structure for its management.

In 2024, there was a concerted effort by bipartisan senators and repre-sentatives to introduce new AI-related legislation. Senators Durbin, Gra-ham, Klobuchar, and Hawley presented the Disrupt Explicit Forged Images and Non-Consensual Edits Act of 2024 (Defiance Act), aimed at holding individuals accountable for disseminating nonconsensual sexual deepfakes. Meanwhile, Congress members Lieu, Nunn, Beyer, and Molinaro proposed the Federal Artificial Intelligence Risk Management Act, which mandated federal and their AI contractors follow best practices in mitigating AI-re-lated risks. Representatives Salazar and Dean put forward the No Artificial Intelligence Fake Replicas And Unauthorized Duplications (No AI Fraud) Act to establish federal protections against AI-generated imitations of in-dividuals' likeness and voices. Additionally, Congress members Eshoo and Beyer, with senators Markey and Heinrich, introduced the Artificial Intel-ligence Environmental Impacts Act of 2024, calling on the National In-stitute of Standards and Technology to create standards for appraising and reporting AI's environmental consequences and encouraging AI developers to voluntarily report those impacts.

As of March 2025, none of the mentioned proposals had gained congres-sional approval, despite multiple appeals to Congress to reach a consensus and establish a federal framework in the United States for AI and data gov-ernance. Currently, the future of comprehensive AI legislation in the US remains uncertain.

5.3.6 How a second presidency for Donald Trump might shape AI policy and regulation

Donald Trump was elected 47th President of the United States in Novem-ber 2024, returning to the White House for a second mandate after one mandate fulfilled by President Biden. When President Trump terminated his first serving, AI had not yet shown all its full potential, and the global framework and scenarios were very different from those we are facing today.

Already during the electoral campaign, President Trump emphasized the importance of reviewing Biden's Executive Order «Safe, Secure, and Trustworthy Development and Use of Artificial Intelligence,» stating his preference for removing regulatory barriers in order to stimulate innovation and strengthen the US leadership in AI, particularly in competition with China. Indeed, the same day of his inauguration, he revoked President Biden's Executive Order, and only a few days later, he signed a new executive order titled «Removing Barriers to American Leadership in Artificial Intelligence,» which seeks to eliminate policies viewed as obstacles to AI innovation. In fact, the order directs federal agencies to identify and revoke regulations that may hinder AI development and mandates the creation of a comprehensive AI action plan to sustain and enhance America's AI dominance.

Central to President Trump's vision is the belief that national security and economic dominance rely on an environment that prioritizes fostering innovation, especially in the AI space. This naturally leads to a more *laissez-faire* regulatory approach, with less emphasis on safety and ethics concerns compared to the approaches of the previous Biden and the Obama administration. Additionally, the Trump administration announced a $500 billion initiative named «Stargate,» in collaboration with companies such as OpenAI, Oracle, and SoftBank. This project aims to develop extensive AI infrastructure and is expected to create over 100,000 jobs in the United States within the next four years.

On another note, the re-election of President Donald Trump has led to shifts in artificial intelligence and fact-checking policies on social media platforms. In fact, as AI algorithms are also being used to scan large volumes of content, identifying posts that may contain misinformation or require human review, they have been seen by the Trump administration as a potential limitation of free speech. Thus, as part of his agenda, President Trump signed an Executive Order titled «Restoring Freedom of Speech and Ending Federal Censorship,» aimed at reducing what the administration considers as excessive intervention by social media platforms and broadcasters in moderating content. These policy changes have immediately led social media companies to reassess their content moderation strategies; a notable development in this area is Meta's (formerly Facebook) decision to discontinue its third-party fact-checking program, which previously relied on independent organizations to verify the accuracy of posts and flag misinformation. Meta introduced a «community notes» system, where users contribute contextual information to potentially misleading posts. This shift has been compared to X's (formerly Twitter) approach and reflects a broader movement toward decentralized content moderation.

Meta's decision, along with broader regulatory changes, underscores the ongoing tension between maintaining an open platform that guarantees freedom of speech and ensuring the accuracy of information online. As artificial intelligence becomes increasingly central to content management, these developments raise crucial questions about the future of policies aimed at tackling online misinformation.

However, when it comes to health issues, the topic of online misinformation may become very delicate, as the experience of Covid-19 pandemic shown.[30] Information not scientifically based in the health space can lead individuals to make harmful decisions and delay necessary treatments or avoid proven interventions. Social media can exacerbate this situation due to the rapid dissemination of such information and the lack of expert oversight. This, in turn, diminishes public trust in official and reliable sources, as clinicians and public health organizations, resulting in increased preventable illnesses and mortality. Therefore, striking a balance between freedom of speech and harmful health content is still crucial.

Last, cybersecurity is likely to remain a top priority for the US administration. This topic has always been a concern for President Trump, who introduced, during its previous mandate, a national cybersecurity strategy with the purposes of defending the homeland by protecting networks, systems, functions, and data and promoting American prosperity by nurturing a secure, thriving digital economy and fostering strong domestic innovation.[31]

5.4 The crucial role of federal agencies in AI regulation

US federal agencies have been instrumental in forming the policy landscape for artificial intelligence, issuing new regulations or guidelines, as well as playing a pivotal role in executing AI governance as mandated by the government. Moreover, as AI has the capability to transform agency operations, it necessitates that these entities concentrate not just on influencing external policies but primarily on refining their own internal procedures to guarantee the ethical application of AI technology.

---

[30] Hye Kyung Kim, and Edson C. Tandoc Jr., «Consequences of Online Misinformation on COVID-19: Two Potential Pathways and Disparity by eHealth Literacy», *Frontiers in Psychology*, *13*, 2022.
[31] White House, National Cyber Strategy of the United States, 2018.

5.4.1 A common effort to develop AI governance

Nearly every federal agency has been involved in shaping the AI policy landscape in the US Among them, National Institute of Standards and Technology (NIST) plays a leading role in developing technical standards for AI technologies. In fact, NIST serves as a neutral convener among industry leaders, researchers, and civil society organizations, fostering collaboration for effective AI regulations. It also engages in international discussions with entities like the EU and OECD to align US AI standards globally, enhancing its role in international technology governance. A key achievement by NIST was the creation of the «US Leadership in AI: A Plan for Federal Engagement in Developing Technical Standards and Related Tools» in August 2019, following President Trump's executive order «Maintaining American Leadership in Artificial Intelligence.» This document set the stage for US governmental AI standards and proposed actions for increasing US leadership in AI. The cornerstone of establishing US federal AI governance, however, is marked by the unveiling of NIST's «AI Risk Management Framework»[32] in January 2023. This framework is designed for discretionary application and aims to enhance trustworthiness in AI technologies through conscientious design, development, deployment, and assessment practices regarding AI products, services, and systems. Following President Biden's Executive Order on AI, NIST was assigned to develop guidelines for safe AI practices across federal agencies. This involves creating industry standards for advanced AI models and establishing testing environments to evaluate these systems. Thus, in July 2024 NIST released the «Artificial Intelligence Risk Management Framework: Generative Artificial Intelligence Profile,»[33] to help organizations identify unique risks posed by generative AI and propose actions for generative AI risk management that best aligns with their goals and priorities.

Another important player in AI policy is the Federal Trade Commission (FTC), which released in April 2020 fresh guidelines[34] concerning the deployment of artificial intelligence and algorithmic systems, following a 2018 session in which the FTC examined AI, algorithms, and predictive analytics. Acknowledging the widespread use of these technologies across various sectors like healthcare, the FTC also cautioned against the inherent

---

[32] National Institute of Standards and Technology, *Artificial Intelligence Risk Management Framework (AI RMF 1.0)*, 2023.

[33] National Institute of Standards and Technology, *NIST-AI-600-1, Artificial Intelligence Risk Management Framework: Generative Artificial Intelligence Profile*, 2024.

[34] Andrew Smith, *Using Artificial Intelligence and Algorithms*, FTC Bureau of Consumer Protection, April 8, 2020.

risks that need to be addressed: For example, one of the examples provided referred to an occasion in which an algorithm, intended to distribute medical services, inadvertently directed more resources towards healthier, predominantly white demographics.

Given that the FTC is authorized to regulate deceptive and unfair business practices under the FTC Act and this authority includes AI technologies, the agency is enabled to address issues such as bias, discrimination, and privacy violations that may result from AI applications. Accordingly, in April 2023, the Federal Trade Commission revisited the issue of discrimination with the collaboration of the Consumer Financial Protection Bureau, the Civil Rights Division of the Justice Department, and the Equal Employment Opportunity Commission. Together, they issued a statement[35] making it clear that their regulatory powers extend to automated systems, which include AI and software designed to automate processes and aid in decision-making or task completion. They identified potential sources of bias and discrimination such as data quality, lack of model transparency, and issues in design and application. Moreover, in May 2023, the US Equal Employment Opportunity Commission published a guidance document titled «Assessing Adverse Impact in Software, Algorithms, and Artificial Intelligence Used in Employment Selection Procedures Under Title VII of the Civil Rights Act of 1964.»

With the aim of encouraging public trust in AI use, the FTC also introduced the Consumer Sentinel Network, an investigative digital tool that grants Sentinel members access to a vast collection of consumer reports, which has been utilized to compile a comprehensive picture of public concerns about AI for more effective FTC response.

Aside from the previously mentioned bodies, many other federal entities have provided guidance on AI governance or led distinct AI initiatives. For instance, the US Patent and Trademark Office, operating under the Department of Commerce, established an AI/emerging technologies partnership[36] designed to more closely analyze the application of such technologies in the areas of patent and trademark assessments and their implications for intellectual property law.

---

[35] Rohit Chopra, Director of the Consumer Financial Protection Bureau, Kristen Clarke, Assistant Attorney General for the Justice Department's Civil Rights Division, Charlotte A. Burrows, Chair of the Equal Employment Opportunity Commission, and Lina M. Khan, Chair of the Federal Trade Commission, *Joint statement on enforcement efforts against discrimination and bias in automated systems*, April 2023.

[36] Patent and Trademark Office, *Events for the Artificial Intelligence and Emerging Technologies Partnership*, 2022.

5.4.2 The Food and Drug Administration (FDA)'s initiatives related to AI regulation

The Food and Drug Administration plays a crucial role in regulating AI in healthcare in the US by ensuring patient safety, adapting to technological advancements, fostering collaboration among stakeholders, and addressing ethical considerations. A review[37] of drug and biologic product regulatory filings with the FDA from 2016 to 2021 revealed a growing trend in the incorporation of artificial intelligence/machine learning (AI/ML): there was only one such filing in each of the years 2016 and 2017; then, there was an annual doubling to tripling in filings from 2018 to 2020; followed by a significant jump to 132 filings in 2021, which is roughly a tenfold increase from the previous year. AI/ML has been applied to various purposes within these submissions, including drug discovery and repurposing, improving clinical trial design, optimizing dosages, enhancing adherence to medication regimens, evaluating endpoints and biomarkers, and conducting post-market monitoring.

Furthermore, the overall number of AI/ML-powered medical device submissions has been on the rise: the FDA cleared the first AI algorithm in 1995, and less than 50 algorithms were approved in the subsequent 18 years. As of 2023, there were over 692 artificial intelligence medical algorithms that have been cleared for use in the US.[38] A significant 87% of devices approved in the year 2022 were for radiology (122), followed by cardiovascular at 7% (10), and 1% each for neurology (2), hematology (1), gastroenterology/urology (1), ophthalmic (2), clinical chemistry (1), and ear, nose, and throat (1).[39] Up to October 19, 2023, no device utilizing generative AI, artificial general intelligence (AGI), or powered by large language models has been authorized,[40] while in 2024 there are nearly 950 authorized AI/ML-enabled medical devices.[41] This delineates a marked distinction between current AI/ML applications in medical devices and the more advanced generative AI

---

[37] Qi Liu, *et al.*, «Landscape analysis of the application of artificial intelligence and machine learning in regulatory submissions for drug development from 2016 to 2021», *Clinical Pharmacology and Therapeutics*, 113(4), 2023, pp. 771-774.

[38] Derek L. G. Hill, «AI in imaging: the regulatory landscape», *The British Journal of Radiology*, 97(1155), 2024, pp. 483-491.

[39] Dave Fornell, *FDA has now cleared 700 AI healthcare algorithms, more than 76% in radiology*, Health Imaging, 2023.

[40] Geeta Joshi, *et al.*, «FDA-approved artificial intelligence and machine learning (AI/ML)-enabled medical devices: an updated landscape», *Electronics* 13(3), 498, 2024.

[41] FDA, Artificial Intelligence and Machine Learning (AI/ML)-Enabled Medical Devices, updated in October 2023.

technologies, which have not yet attained FDA clearance. The authorized devices predominantly employ conventional AI and machine learning techniques rather than generative AI or Artificial General Intelligence (AGI). The FDA has underscored that although there is considerable interest in these cutting-edge technologies, they have not yet been incorporated into approved medical devices.

As AI/ML continues to be investigated for its potential to expedite drug development and power medical devices, the FDA is faced with the need to tackle emerging safety and efficacy concerns, along with issues related to unauthorized data sharing and cybersecurity threats. Keeping these factors in view, several initiatives have been launched in recent years.

The CDRH Digital Health Center of Excellence was established in 2020 within the FDA's Center for Devices and Radiological Health, with the aim of advancing digital health technologies within the context of the FDA's regulatory responsibilities. In January 2021, the Center introduced the «AI/ML-Based Software as a Medical Device Action Plan,» which set forth the FDA's strategy for supervising these innovative tools. This plan came as an answer to the feedback provided by stakeholders to the April 2019 discussion paper entitled «Proposed Regulatory Framework for Modifications to Artificial Intelligence/Machine Learning-Based Software as a Medical Device.» It proposed five key initiatives: further developing the proposed regulatory framework, including through the issuance of draft guidance on a predetermined change control plan (for software's learning over time); supporting the development of good machine learning practices to evaluate and improve machine learning algorithms; fostering a patient-centered approach, including device transparency to users; developing methods to evaluate and improve machine learning algorithms; and advancing real-world performance monitoring pilots.

Following the White House's release of its Blueprint for an AI Bill of Rights on October 4, 2022, the FDA issued its first draft guidance in April 2023, titled «Marketing Submission Recommendations for a Predetermined Change Control Plan for Artificial Intelligence/Machine Learning-Enabled Device Software Functions» (PCCP Draft Guidance). This guidance aimed to advance machine learning-enabled device software functions by allowing iterative updates without constant FDA re-approval, thus promoting innovation while ensuring safety. The authority to approve such PCCPs was granted to the FDA by Congress in the Food and Drug Omnibus Reform Act of 2022.

In May 2023, the FDA published a discussion paper titled «Using Artificial Intelligence and Machine Learning in the Development of Drug and Biological Products.» The paper advocates for a risk-based approach when integrating AI/ML into drug innovation and public health protection. It offers insights into current and prospective applications of AI/ML in drug development, discusses potential risks and ways to mitigate them, and highlights the role of human oversight in technology implementation. In conjunction with this, the Center for Drug Evaluation and Research (CDER) put forth another discussion paper, «Artificial Intelligence in Drug Manufacturing,» under the Framework for Regulatory Advanced Manufacturing Evaluation (FRAME) Initiative, to address AI's incorporation in pharmaceutical production. On March 15, 2024, the FDA issued another paper titled «Artificial Intelligence and Medical Products: How CBER, CDER, CDRH, and OCP are Working Together,» outlining a coordinated approach across various FDA centers to regulate AI in medical products. Furthermore, on October 11th, 2023, the FDA announced the establishment of the Digital Health Advisory Committee. This committee is set to provide guidance to the Food and Drugs Commissioner on digital health matters, including artificial intelligence, bringing valuable knowledge and perspective to bolster the FDA's grasp of digital technologies' advantages, possible risks, and clinical impacts.

In January 2025, the FDA released the «Considerations for the Use of Artificial Intelligence to Support Regulatory Decision-Making for Drug and Biological Products Guidance for Industry and Other Interested Parties» draft guidance,[42] providing recommendations on how the FDA plans to apply a risk-based credibility assessment framework to evaluate the use of artificial intelligence models that produce information or data intended to support regulatory decision-making regarding safety, effectiveness, or quality of drugs and biological products. Also, in the same timeframe, draft guidance on «Artificial Intelligence-Enabled Device Software Functions: Lifecycle Management and Marketing Submission Recommendations»[43] was released, discussing recommendations on the contents of marketing submissions for devices that include AI-enabled device software functions including documentation and information that will support FDA's review, as well as for the design and development of AI-enabled devices that manufacturers may consider using throughout the TPLC.

---

[42] FDA, *Considerations for the Use of Artificial Intelligence to Support Regulatory Decision-Making for Drug and Biological Products Guidance for Industry and Other Interested Parties draft guidance*, 2025.
[43] FDA, *Artificial Intelligence-Enabled Device Software Functions: Lifecycle Management and Marketing Submission Recommendations—Draft Guidance for Industry and Food and Drug Administration Staff*, 2025.

5.4.3 Adapting existing frameworks to AI applications

Crafting new regulatory measures entails a considerable investment of time and energy. Yet, it would be overly simplistic to assume that artificial intelligence had been operating before without any regulations. Many current laws, particularly those concerning privacy and liability, are already relevant to AI. Prior to introducing new regulations that may lead to ambiguous overlaps, regulators should always thoroughly evaluate whether the present rules governing medical device regulation adequately fit the purpose.

Policymakers are currently faced with the challenge of understanding how to interpret existing laws in the context of AI. For instance, in their joint statement,[44] the Consumer Financial Protection Bureau, the Department of Justice's Civil Rights Division, the Equal Employment Opportunity Commission, and the Federal Trade Commission stressed how «existing legal authorities apply to the use of automated systems and innovative new technologies just as they apply to other practices.» They expressed worries over «potentially harmful uses of automated systems» and emphasized that they would work «to ensure that these rapidly evolving automated systems are developed and used in a manner consistent with federal laws.» Moreover, the FTC has emphatically reasserted[45] that AI developers and users are already subject to compliance with three regulatory statutes: Section 5 of the FTC Act forbidding unfair or deceptive practices explicitly encompasses scenarios such as marketing or deploying algorithms with racial bias; the Fair Credit Reporting Act is relevant in instances where an algorithm's application leads to the denial of employment, housing, credit, insurance, or other benefits; the Equal Credit Opportunity Act prohibits the use of discriminatory algorithms by lenders that could lead to unequal treatment based on race, color, religion, national origin, gender, marital status, age, or because an individual receives public assistance.

Other examples of existing regulations that govern AI and were not developed to be AI-specific include data governance and privacy laws, discrimination laws, general tort principles and product liability laws, financial regulations, and cybersecurity standards.

---

[44] Rohit Chopra, Director of the Consumer Financial Protection Bureau, Kristen Clarke, Assistant Attorney General for the Justice Department's Civil Rights Division, Charlotte A. Burrows, Chair of the Equal Employment Opportunity Commission, and Lina M. Khan, Chair of the Federal Trade Commission, *Joint statement on enforcement efforts against discrimination and bias in automated systems*, 2023.

[45] Elisa Jillson, *Aiming for truth, fairness, and equity in your company's use of AI*, Federal Trade Commission, Business Blog, April 19, 2021.

5.5 In the wait for federal regulation, States started legislating specific AI uses

In the absence of a federal framework, states began developing their own AI regulations. By the end of 2023, a total of seventeen US states (California, Colorado, Connecticut, Delaware, Illinois, Indiana, Iowa, Louisiana, Maryland, Montana, New York, Oregon, Tennessee, Texas, Vermont, Virginia, and Washington) had passed 29 pieces of legislation aimed at governing the creation or application of artificial intelligence in different fields.[46]

In fact, the absence of specific federal AI legislation and of a national standard leaves policy design to individual states, which have consequently the autonomy to implement and enforce their own laws, resulting in a patchwork of different regulations. Moreover, typically, state legislation targets specific goals rather than establishing a holistic regulatory structure; for example, California implemented the Bolstering Online Transparency (BOT) Act in July 2019, which prohibits individuals or entities from utilizing bots to engage with people in California for promoting sales or influencing electoral votes without revealing the bot's presence.

One of the most pressing concerns recognized by many state officials is to deepen the institutional understanding of AI and enhance their regulatory capabilities. To address this, at least a dozen states, such as Alabama, California, Colorado, Connecticut, Illinois, Louisiana, New Jersey, New York, North Dakota, Texas, Vermont, and Washington,[47] have passed legislation establishing task forces, offices, or councils, or have directed other government-related bodies to study AI policies and their effects. This ap-

---

[46] Lawrence Norden, *States Take the Lead on Regulating Artificial Intelligence*, Brennan Center for Justice, 2023.

[47] AL SB78, Technology, Alabama Council on Advanced Technology, estab., to advise Governor and Legislature, members, duties; CA AB485, Local government: economic development subsidies; CO SB113, Artificial Intelligence Facial Recognition; CT SB01103, An Act Concerning Artificial Intelligence, Automated Decision-making And Personal Data Privacy; IL HB0645, Future of Work Task Force; LA SCR49, Requests the Joint Committee on Technology and Cybersecurity to study the impact of artificial intelligence in operations, procurement, and policy; NJ S2723; 21st Century Integrated Digital Experience Act; NY S03971, Creates a temporary state commission to study and investigate how to regulate artificial intelligence, robotics and automation; and repeals such commission; ND HB1003, Matching grants for legal education and the workforce education advisory council; to provide for a transfer; to provide for a legislative management study; to provide loan authorization for the Mayville state university old main project; to provide for a report; to provide an exemption; to provide legislative intent; and to declare an emergency; TX HB2060, Relating to the creation of the artificial intelligence advisory council; VT H0410, An act relating to the use and oversight of artificial intelligence in State government; WA SB6544, Establishing the future of work task force.

proach reflects the States' ambition to position themselves as leaders in AI technology. For example, in June 2024, New York Governor Kathy Hochul announced the formation of the new Emerging Technology Advisory Board made up of prominent individuals from the tech sector, tasked with developing an advanced technology strategy for the state. Moreover, during the Fiscal Year 2025 budget discussions, Governor Hochul secured a significant agreement with the legislature to found Empire AI, a consortium aiming to propel New York to the forefront of artificial intelligence, bolstered by a $275 million investment.

A further matter addressed by multiple state AI regulations is the AI discrimination risk. States like California, Colorado, and Connecticut[48] have implemented laws to prevent AI from reinforcing biases against protected classes, ensuring that these systems are developed equitably. Given that algorithmic discrimination poses a significant concern in the healthcare industry, particularly regarding access to care, it's worth discussing both the issue and the proposed remedies.

Illinois was the first state to impose restrictions in 2019, marking a significant advancement in the fight against AI discrimination in the workplace.[49] Specifically, the AI Video Interview Act, which experienced revisions in 2021, mandates that employers utilizing AI-driven evaluations must advise applicants prior to the interview of the potential use of AI in assessing their video interviews and gauging their suitability for the role. They must also provide details before the interview regarding how the AI operates and what categories of attributes it assesses when reviewing applicants. Moreover, employers are required to secure consent from the candidates before the interview to be assessed by the AI as outlined in the provided information.

In 2020, Maryland legislation[50] barred employers from employing facial recognition technology to generate a facial template during a prospective employee's pre-employment interview without obtaining the applicant's consent through a signed waiver.

Additionally, the New York City Council has passed a bill[51] mandating a bias audit on any automated employment decision tool before it is utilized. This legislation also stipulates that candidates or employees living in the city

---

[48] CA AB331, Automated decision tools; CO SB169, Restrict Insurers' Use of External Consumer Data; CT SB01103, An Act Concerning Artificial Intelligence, Automated Decision-making and Personal Data Privacy.

[49] IL HB2557, Video Interview Act.

[50] MD HB1202, Labor and Employment—Use of Facial Recognition Services—Prohibition.

[51] Law 2021/144, A Local Law to amend the administrative code of the city of New York, in relation to automated employment decision tools.

must be informed about the implementation of these tools in their evaluation for employment or promotion and must be provided details regarding the job qualifications and traits that the automated tool will assess.

The last significant issue that was addressed by several state legislations, such as California, Texas, Minnesota, and Washington,[52] is the one regarding AI's influence on elections. Thus, these states have enacted legislation to prohibit or require transparency of media alteration that could mislead voters about political candidates or sway election results.

5.6 Ensuring privacy protection in the US; concerns about an outdated regulatory framework

Effective and ethical regulation of artificial intelligence can only be attained within a framework that also addresses privacy and data governance. As AI technologies rely significantly on large datasets, often including personal information, the relationship between privacy laws and AI regulation is crucial. Additionally, considering the sensitive nature of health data and its importance in training AI systems, it is important to assess whether current data protection laws are sufficient to support the AI industry while protecting personal rights.

Overall, there are significant differences between US and EU data privacy rules. Numerous federal and state laws in the United States address various aspects of data privacy, such as financial information, health-related data, and minors' data. The lack of a recent, comprehensive federal privacy framework has resulted in a collection of sector-specific laws, causing gaps in protections and consumer confusion regarding their rights. There is ongoing political debate and disagreement about the details of a potential federal privacy law, including issues related to the right of preemption (whether federal law would override state laws) and the balance between consumer protection and business interests. On one side, states like California which have already implemented robust privacy laws, are concerned that a federal law might reduce these protections or lower the standards in the country. For instance, the California Privacy Protection Agency has expressed worries

---

[52] CA AB972, Elections: deceptive audio or visual media; TX SB751, Relating to the creation of a criminal offense for fabricating a deceptive video with intent to influence the outcome of an election; MN HF1370, Cause of action for nonconsensual dissemination of deepfake sexual images established, crime of using deepfake technology to influence an election established, and crime for nonconsensual dissemination of deepfake sexual images established; WA SB5152, Defining synthetic media in campaigns for elective office, and providing relief for candidates and campaigns.

that a federal law may weaken the stringent protection established by voters in their state.[53] On the other hand, other stakeholders believe that existing sector-specific laws already provide adequate protection for certain types of data, leading to a lack of urgency for comprehensive federal legislation.

5.6.1 Privacy laws at the federal level

At the national level, the Privacy Act of 1974 is a critical piece of legislation to consider. It manages how federal agencies collect, store, utilize, and share personally identifiable information in their record systems. The act restricts these agencies from releasing personal data without the individual's written consent, except for certain exemptions such as providing data to the Census Bureau for statistical usage. People have the right to access their own records, seek amendments to inaccurate or incomplete records, and are entitled to protection from unjustified privacy intrusions.

In 1996, the Health Insurance Portability and Accountability Act (HIPAA) was enacted by President Clinton. This legislation requires covered entities to honor an individual's right to access and amend their health information and mandates that the use or disclosure of health information is prohibited without the individual's express written consent. The HIPAA establishes three primary rules: the Privacy Rule, which restricts the allowed uses and disclosures of health information by covered entities and business associates without authorization; the Security Rule, which sets standards for managing protected health information (PHI) electronically and physically, ensuring the confidentiality, integrity, and availability of patient data; and the Breach Notification Rule, which mandates the timely reporting of any breaches of PHI.

Additionally, at the federal level, privacy protections include the Gramm-Leach-Bliley Act (GLBA), established in 1998, which regulates data privacy for financial institutions, and the Children's Online Privacy Protection Act (COPPA), also passed in 1998, setting forth restrictions on the handling of data collected from children under the age of 13.

Last, the Biden administration and Congress, worried by the potential of the most recent technological developments and their threat to national biosecurity, have chosen to focus their attention on restricting transactions, data transfers, and certain types of contracts with «countries of concern.» Legislative developments that took place in 2024 include:

---

[53] California Privacy Protection Agency, *The California Privacy Protection Agency Opposes the American Privacy Rights Act*, 2024.

- President Biden's Executive Order «Preventing Access to Americans' Bulk Sensitive Personal Data and United States Government-Related Data by Countries of Concern,» that directs the US Department of Justice (DOJ) to promulgate regulations that restrict or prohibit transactions involving certain bulk sensitive personal data or United States Government-related data and «countries of concern» or «covered persons.» Sensitive personal data include covered personal identifiers, personal financial data, personal health data, precise geolocation data, biometric identifiers, and human genomic data. Also, the DOJ's initially identified countries of concern are People Republic of China (including Hong Kong and Macau), Russia, Iran, North Korea, Cuba, and Venezuela.
- The enactment of the Protecting Americans' Data from Foreign Adversaries Act (PADFA), which makes it unlawful for a «data broker» to sell, license, rent, trade, transfer, release, disclose, provide access to, or otherwise make available the personally identifiable sensitive data of a United States individual to People Republic of China, other countries of concern (Russia, North Korea, and Iran), or any entity «controlled» by those countries. Unlike the DOJ ANPRM, the law does not specifically include Hong Kong or Macau and does not include Cuba or Venezuela.
- The enactment of the Protecting Americans from Foreign Adversary Controlled Applications Act (TikTok Divestment Law), which prohibits marketplaces (including online mobile application stores) and internet hosting services in the United States from supporting certain applications owned by a foreign adversary. However, there is speculation that the Trump administration may revisit the ban, potentially reversing or modifying the current restrictions.

5.6.2 The HIPAA leaving space for health data protection concerns

An average extensive electronic health record is estimated to be valued at around \$250.[54] Nevertheless, if combined with genetic information, its worth could potentially exceed \$6,500.[55] Still, this evaluation only skims the surface of the true value of health data, which has generally been limited to short-term clinical information collected in hospital and medical office

---

[54] Trustwave, *2018 Global Security Report*, 2018.
[55] EY, *Realising the Value of healthcare Data: A Framework for the Future*, 2019.

settings. Within this framework, one can appreciate the actual worth of data within the healthcare sector.

In contrast with the European Union, the healthcare industry in the United States is less regulated, offering improved access to medical data through open data policies by the government and efforts from the private sector. However, this poses privacy issues owing to inadequate legal safeguards for anonymized data, which may be vulnerable to re-identification in the future.[56]

Indeed, HIPAA was enacted nearly three decades ago, in 1996, way before smartphones, algorithms, and AI, and it doesn't provide adequate protection against the challenges posed by new technologies. One primary limitation of data protection under HIPAA is its application only to designated «covered entities,» which comprise healthcare providers (such as doctors, nurses, psychologists, and dentists), health plans (including health insurance companies and government programs like Medicare), and healthcare clearinghouses that process medical data. Consequently, any health information shared with other parties (like schools or employers) or through different channels (such as social media or certain applications) falls outside the scope of HIPAA's protective measures. Furthermore, another limit to data protection pertains to the use and dissemination of protected health information for purposes such as medical research, policy evaluation, and other studies without needing individual permission if the data has been de-identified. This implies that healthcare organizations can capitalize on their extensive data resources: They can strip patient names, locations, contact numbers, and other personal identifiers from records, thereafter freely distributing or selling the anonymized datasets to research collaborators. In fact, once data is de-identified according to HIPAA standards, it is no longer considered protected health information (PHI) and is not subject to HIPAA regulations. Thus, during this process, there is no obligation to acquire patient consent or notify them about the data-sharing activities, allowing organizations to use the data effectively while remaining compliant with privacy standards set by HIPAA. Nevertheless, AI algorithms can now collect and analyze health data from various sources, and once de-identified data is combined with other datasets, the danger of re-identification becomes

---

[56] On the topic, see, between the others: Mehri Sadri, «HIPAA: a demand to modernize health legislation», *The Undergraduate Law Review at UC San Diego*, 2(1), 2024; Kim Theodos, and Scott Sittig, «Health information privacy laws in the digital age: HIPAA doesn't apply», *Perspectives in health information management* vol. 18, Winter 11, 2020; Francesca A. Sacchi, *Healthcare data: a new currency was born. Examples from Israel and the United States*, Medialaws.eu, July 25, 2023.

significant. Presently, the main deterrence against this risk lies solely in an agreement from data recipients to refrain from attempting re-identification.[57] However, these agreements may not be legally binding or enforceable in all cases, raising concerns about their effectiveness.

Regrettably, many consumers may mistakenly assume that all their health information is protected under HIPAA. However, much of the data gathered outside traditional healthcare settings does not fall under this protection and this misconception may result in a lack of awareness regarding the privacy risks posed by new technologies. Moreover, in this era of widespread artificial intelligence, digital platforms, and social networks, the protection provided by HIPAA is somewhat restricted, especially because much of the health information exchange occurs outside its jurisdiction. Additionally, there are significant privacy worries due to the possibility that de-identified data might be re-identified, which compromises patients' ability to manage their personal health information effectively.

5.6.3 California privacy laws raising the US standards once again

California, home to Silicon Valley, is a hub of tech innovation and AI development. The high concentration of talent and investment has underscored the need for regulatory frameworks to address ethical and societal impacts. Consequently, California legislators are working to establish guidelines for responsible innovation and public protection. Notably proactive in AI and privacy laws, California responds to public concerns about risks like algorithmic bias, data sharing issues, and job loss by promoting its own state initiatives. With no federal framework yet in place, California's actions and framework serve indeed as a model for other US states, helping shape national standards and practices in AI governance.[58]

In general, California stands out for enacting the most rigorous privacy legislation in the United States. In fact, since Congress enacted HIPAA, it explicitly allowed for the enforcement of stronger state health privacy laws; California has thus enacted state laws under this provision with the intent to offer enhanced protection.

---

[57] C. Christine Porter, «Constitutional and regulatory: de-identified data and third party data mining: the risk of re-identification of personal information», in *Shidler Journal of Law, Commerce & Technology*, 2008.

[58] For example, see Stuart L. Pardau, «The California Consumer Privacy Act: towards a European-style privacy regime in the United States», *Journal of Technology Law & Policy*, 23(68), 2018–2019; Catherine Barrett, «Are the EU GDPR and the California CCPA becoming the de facto global standards for data privacy and protection?», *Scitech Lawyer*, 15 (Fasc. 3 Spring), 2019, pp. 24-29.

Already in 2009, the Health Information Technology for Economic and Clinical Health Act (HITECH), as part of the American Recovery and Reinvestment Act (ARRA), led California to broaden the scope of HIPAA privacy and security rules. This expansion increased liability and enforcement actions for non-compliance.

In 2018, the state passed the Consumer Privacy Act (CCPA), enhancing consumer powers over their data with regard to business usage; Californians ensured significant privacy rights such as the ability to understand the scope of personal information gathered and its intended use or distribution; the right to erase certain personal data a business holds; the option to refuse the sale or transfer of their personal data; and protection against discrimination for availing themselves of their CCPA entitlements.

Building on these provisions, voters approved the California Privacy Rights Act (CPRA) in November 2020, which further extended these protections from January 1, 2023, by allowing citizens the right to correct inaccurate personal information and limit how businesses handle and divulge sensitive personal details. The CPRA strengthens requirements, aligns California more with the GDPR, and establishes the California Privacy Protection Agency. Previously managed by the attorney general, consumer privacy now has a dedicated agency and budget. This agency will also receive part of the fines and settlements from companies that violate the law.

Last, effective in 2022, the Confidentiality of Medical Information Act (CMIA) is a California statute that supplements HIPAA by safeguarding personally identifiable medical information. Under the Act, healthcare providers, service plans, or contractors cannot disclose patient information without authorization unless an exception applies. Also, entities such as healthcare providers, service plans, pharmaceutical companies, and contractors must protect the confidentiality of medical records during creation, storage, maintenance, destruction, or disposal. Last, medical information includes data related to a patient's medical history, mental or physical condition, and treatment, whether in electronic or physical form.

Given California's significant advancements in privacy protections, it is understandable that California lawmakers and privacy advocates express concerns over a federal privacy law potentially preempting state laws and thereby weakening existing safeguards. This results in hesitation to fully support a federal privacy solution without guarantees that state-level protections will be preserved or improved. However, the California Consumer Privacy Act (CCPA) and its subsequent amendment, the California Privacy Rights Act (CPRA), have served as models for other states developing their own privacy legislation. States such as Virginia, Colorado, Connecticut, and Utah, among others, have introduced laws that reflect some aspects of California's approach to consumer data protection. As of 2024, numerous states are either enacting or considering privacy laws that incorporate principles found in California's legislation, underscoring its influence as a leading example in the national discourse on data privacy.

5.6.4 State scenarios in evolution

As artificial intelligence expands and concerns over privacy increase among the population, recent years have seen an uptick in the introduction and passage of extensive privacy legislation at the state level. As of 2023, the United States saw the activation of additional privacy frameworks that have broadened the landscape of 13 state-specific privacy regulations. Specifically, states like Virginia, Colorado, Connecticut, and Utah took inspiration from California laws in developing their own legislations that still result quite similar to each other even if they were tailored to fit their specific regulatory landscape and consumer needs. In particular:

- The Virginia Consumer Data Protection Act (VCDPA) became effective on January 1, 2023. While it addresses consumer rights related to data processing, it does not specifically introduce guidelines for data profiling and automated decision-making processes in the same way that California does. However, it does allow consumers to opt out of targeted advertising and profiling.
- The Colorado Privacy Act (CPA), which went into force on July 1, 2023, also grants consumers the ability to opt-out of profiling in circumstances where it is related to automated decision-making with potential legal or similarly significant ramifications.
- The Connecticut Privacy Act (CTPA), which simultaneously went into force on July 1, 2023, presents many of the law's provisions reflecting a blend of the Virginia and Colorado statutes.
- The Utah Consumer Privacy Act, which went into force on December 31, 2023, is closely inspired by the Virginia Consumer Data Protection format while also echoing elements from the California Consumer Privacy Act. However, the Act adopted a more business-friendly stance compared to its antecedents in the realm of consumer privacy legislation.

In 2024, the Oregon Consumer Privacy Act, the Texas Data Privacy and Security Act (TDPSA), and the Montana Consumer Data Privacy Act became effective. The TDPSA aligns with privacy laws in other states like Virginia and Colorado but its approach is usually considered more market-driven when compared to California's more stringent regulations. The year 2025 is set to see the enactment of the Delaware Personal Data Privacy Act, Iowa Consumer Data Protection Act, New Jersey SB 332, and Tennessee Information Protection Act. Following these, the Indiana Consumer Data Protection Act is anticipated to take effect in 2026.

6 Local Regulations for Local Needs: A Global Perspective on AI Regulation

6.1 The adoption of the risk-based approach worldwide: current practices in the G7 countries beyond EU and US

The risk-based approach, adopted by the European Union and endorsed by the United States, is often seen as the optimal method for balancing risk management with innovation protection.[1] Consequently, it is rapidly being adopted worldwide, though it may be tailored locally according to each country's needs.

Already in April 2023, the G7 Digital and Tech Ministers' Meeting released a Ministerial declaration in which they «reassert that AI policies and regulations should be risk-based and forward-looking to preserve an open and enabling environment for AI development and deployment that maximises the benefits of the technology for people and the planet while mitigating its risks.»[2]

A significant milestone was also the AI Safety Summit 2023, organized by the United Kingdom and conducted in November 2023 at Bletchley Park in Buckinghamshire.[3] This event gathered international governments, top AI companies, civil society organizations, and research experts to examine the risks associated with advanced AI development and to deliberate on how these risks can be mitigated through globally coordinated efforts. The first output of the meeting was «The Bletchley Declaration on AI safety,»[4] signed by 28 countries (including the European Union, United States, China, Kingdom of Saudi Arabia, United Arab Emirates, India, Indonesia, The Philippines, Republic of Korea, Kenya, Nigeria, and Rwanda) which reinforced the principle of using a risk-based approach, stating that «many risks

---

[1] Raphaël Gellert, «The role of the risk-based approach in the General data protection Regulation and in the European Commission's proposed Artificial Intelligence Act: business as usual?», *Journal of Ethics and Legal Technologies*, 3(2), 2021.

[2] Ministerial Declaration, The G7 Digital and Tech Ministers' Meeting, April 30, 2023,

[3] All information regarding the Summit is provided on the designated webpage specifically created by the UK government at https://www.gov.uk/government/topical-events/ai-safety-summit-2023.

[4] Countries Attending the AI Safety Summit, The Bletchley Declaration, 1st-November 2, 2023.

arising from AI are inherently international in nature, and so are best addressed through international cooperation. We resolve to work together in an inclusive manner to ensure human-centric, trustworthy and responsible AI that is safe and supports the good of all through existing international fora and other relevant initiatives, to promote cooperation to address the broad range of risks posed by AI. In doing so, we recognise that countries should consider the importance of a pro-innovation and proportionate governance and regulatory approach that maximises the benefits and takes into account the risks associated with AI. This could include making, where appropriate, classifications and categorisations of risk based on national circumstances and applicable legal frameworks.» Then, the Declaration goes on indicating that «in the context of our cooperation, and to inform action at the national and international levels, our agenda for addressing frontier AI risk will focus on: (1) identifying AI safety risks of shared concern, building a shared scientific and evidence-based understanding of these risks, and sustaining that understanding as capabilities continue to increase, in the context of a wider global approach to understanding the impact of AI in our societies; (2) building respective risk-based policies across our countries to ensure safety in light of such risks, collaborating as appropriate while recognising our approaches may differ based on national circumstances and applicable legal frameworks. This includes, alongside increased transparency by private actors developing frontier AI capabilities, appropriate evaluation metrics, tools for safety testing, and developing relevant public sector capability and scientific research.» The countries also agreed to support the development of an international, independent, and inclusive «State of the Science» Report on the capabilities and risks of frontier AI.

Another significant outcome from the Summit was the joint announcement on AI safety testing by global leaders (including Australia, Canada, the European Union, France, Germany, Italy, Japan, the Republic of Korea, Singapore, the United States of America, and the United Kingdom) and companies developing AI systems (like Amazon Web Services, Anthropic, Google, Google DeepMind, Inflection AI, Meta, Microsoft, Mistral AI, Open AI, and xAI).[5] They expressed a collective aim to bolster public confidence in AI safety, starting with an increased focus on AI safety testing and research. This joint announcement marked a crucial development in international policymaking, highlighting the essential involvement of companies in the process, allowing them to issue statements jointly with global leaders.

---

[5] *Safety Testing: Chair's Statement of Session Outcomes*, November 2, 2023.

In May 2024, Australia, Canada, the European Union, France, Germany, Italy, Japan, the Republic of Korea, the Republic of Singapore, the United States of America, and the United Kingdom signed the Seoul Declaration for safe, innovative, and inclusive AI and Seoul Statement of Intent toward International Cooperation on AI Safety Science, [6] which, building upon the Bletchley Declaration, and aimed to promote safe, secure, and trustworthy AI development, addressed global challenges and the protection of human rights, and committed nations to create AI safety institutes and work together on AI safety science, sharing information about AI models, risks, and safety incidents.

Overall, many nations are now adopting a risk-based strategy, beginning with the G7 countries (Canada, France, Germany, Italy, Japan, the United Kingdom, and the United States). While Canada's framework is heavily influenced by that of the EU (though it targets only high-risk systems) the United Kingdom aligns more closely with the US, aiming not to stifle innovation and choosing a sector-specific approach over a broad, horizontal framework. Meanwhile, Japan is concentrating more on establishing an AI environment grounded in ethical principles.

6.1.1 Canada

Canada is planning to regulate AI with the Artificial Intelligence and Data Act (AIDA), proposed in 2021 and introduced by the Government of Canada as part of Bill C-27, the Digital Charter Implementation Act, in June 2022. Unlike the EU, the AIDA focuses solely on high-risk systems. Additionally, a high-risk system is not defined by a single definition but rather by a list of key factors including: evidence of risks to health and safety or potential adverse impacts on human rights; severity of possible harms; scale of usage; nature of harms or negative impacts that have already occurred; practicality or legality of opting out of such a system; economic or social imbalances, or age of affected individuals; and the adequacy of regulation under other laws. An explicit example of high-risk model in health is made by quoting certain AI applications that are «integrated in health and safety functions, for example making critical decisions or recommendations on the basis of data collected from sensors. These include autonomous driving systems and systems making triage decisions in the health sector. These AI systems have the potential to cause direct physical harm, while bias may also result if risks have not been ad-

---

[6] Seoul Declaration and Statement of Intent toward International Cooperation on AI Safety Science, agreed by world leaders represented at the AI Seoul Summit, May 21, 2024.

equately mitigated.»[7] The Act also designates the AI and Data Commissioner as the regulator and enforcer, introducing new criminal penalties to limit AI uses that result in significant harm. These penalties include administrative monetary fines, prosecution of regulatory offenses, or a separate system for actual criminal offenses. Amendments made to the text in November 2023 were aimed at aligning the AIDA with the European AI Act.

Canada's federal government institutions must also adhere to the Directive on Automated Decision-Making, which aims to ensure that AI-driven administrative decisions align with fundamental principles of administrative law such as transparency, accountability, legality, and procedural fairness. This directive employs a risk-based approach, categorizing AI into risk levels: low, moderate, high, and very high.

Regarding generative AI, additional initiatives are in progress. These include a set of principles for the responsible development and use of generative AI established by Canadian privacy regulators (comprising federal, provincial, and territorial privacy authorities), as well as a voluntary code of practice for generative AI.

Current consumer protection regulators are also taking steps to manage some of the effects of AI under their legal frameworks. For example, Health Canada has issued in 2023 guiding principles for the development of medical devices that use machine learning. The 10 principles cover multiple aspects, asking to leverage multidisciplinary expertise throughout the total product life cycle to promote a better understanding of a model's intended integration into clinical workflow, and the desired benefits and associated patient risks, or to implement good software engineering and security practices. The principles also focus on diversity, as they require that clinical study participants and datasets are representative of the intended patient population, while training datasets should be independent of test sets and selected reference datasets based upon the best available methods. The other principles are that the model design is tailored to the available data and reflects the intended use of the device; the focus is placed on the performance of the human-AI team (so, where the model has a «human in the loop,» human factors considerations and the human interpretability of the model outputs are addressed with emphasis on the performance of the human-AI team, rather than just the performance of the model in isolation); the testing demonstrates device performance during clinically relevant conditions; the users are provided clear, essential information; and the deployed models are monitored for performance and retraining risks are managed.

---

[7] Government of Canada, The Artificial Intelligence and Data Act (AIDA)—Companion document, March 13, 2023.

Also, in October 2024, the Office of the Privacy Commissioner of Canada (OPC) released a statement[8] in response to the Standing Committee on Access to Information, Privacy and Ethics (ETHI) report regarding the federal government's use of technological tools to extract personal data from mobile devices and computers. The OPC expressed support for the report's findings and highlighted the critical need to update the Privacy Act to address the changing digital environment, recommending that privacy law reforms should mandate federal institutions to conduct privacy impact assessments before using high-risk tools (AI or otherwise), integrate privacy by design into new technologies, consult the OPC prior to launching privacy-sensitive initiatives, and ensure greater transparency and accountability in government operations.

6.1.2 Japan

Japan, renowned for its dynamic robotics industry, has significantly influenced the G7 agenda by initiating the Hiroshima AI process[9] under Japan's presidency in May 2023, with the aim of promoting safe, secure, and reliable AI. However, Japan has yet to approve or create a specific law for AI, as the country is currently concentrating on developing an ethical, principle-based framework.

Based on instructions issued by the Prime Minister in «Public-Private Dialogue towards Investment for the Future» in April 2016, the national government established the «Strategic Council for AI Technology,» created to develop and steer the country's approach to artificial intelligence, ensuring Japan stays competitive in the fast-evolving AI sector.[10] The council was part of a broader initiative to integrate AI technologies into Japan's economic and industrial strategies, while also addressing related ethical, social, and regulatory issues. The Council, acting as a control tower, manages five National Research and Development Agencies that fall under the jurisdiction of the Ministry of Internal Affairs and Communications, Ministry of Education, Culture, Sports, Science and Technology, and Ministry of Economy, Trade and Industry. In addition to promoting research and development of

---

[8] Statement by the Privacy Commissioner following release of committee report on the federal government's use of technological tools capable of extracting personal data from mobile devices and computers, October 11, 2024.

[9] Further information can be found on «The Hiroshima AI Process: Leading the Global Challenge to Shape Inclusive Governance for Generative AI» Japan Government website, February 9, 2024.

[10] Strategic Council for AI Technology, *Artificial Intelligence Technology Strategy— Report of Strategic Council for AI Technology*, March 31, 2017.

AI technology, the Council coordinates with industries related to the industries that utilize AI (so-called «exit industries»), and is moving forward with social implementation of AI technology.

In 2019, Japan introduced the «Social Principles of Human-Centric AI,»[11] which include human-centric approaches, education and literacy, privacy protection, security assurance, fair competition, fairness, accountability, transparency, innovative systems, and governance, with the purposes of mitigating potential negative impacts while leveraging AI to benefit society. The aim of the government was, in fact, to promote a transition to an «AI-Ready Society,» where AI can be utilized effectively and safely and can impact various aspects of society, as human potential, social systems, industrial structures, and innovation. In 2022, Japan's Ministry of Economy, Trade and Industry released guidelines to implement these principles.[12]

In April 2023, the Digital Society Promotion Headquarters of the Liberal Democratic Party and its Project Team on Evolution and Implementation of AI released a whitepaper titled «Japan's National Strategy in the New Era of AI.»[13] The whitepaper raised concerns about high-risk AI systems and recommended analyzing AI regulations in foreign countries like the EU, the United States, and China to identify areas and sectors in which legal measures or regulations may be necessary. Significant violations of human rights, national security issues, and interference with the democratic process were highlighted as the most sensitive areas. The whitepaper also suggested that Japan should play an active and strategic role in international discussions on AI rules through various forums, including the G7 Summit, and collaborate with other nations to create global frameworks for AI use. Additionally, it proposed that existing regulations be adapted flexibly to the new AI era by enhancing the speed and user-friendliness of current regulatory reform procedures like the Regulatory Reform Council, regulatory sandbox, and gray zone improvement process to develop an environment where businesses can explore new ventures without being constrained by existing regulations.

Japan, as other countries, has been currently seeking companies' support in developing and implementing rules for safe AI use, opening a dialogue with multiple stakeholders to collect different perspectives. For example, the Japanese branches of major tech companies, including Microsoft, Google,

---

[11] Available at https://www.cas.go.jp/jp/seisaku/jinkouchinou/pdf/humancentric-ai.pdf.

[12] Release of Japan's Guidelines on Respecting Human Rights in Responsible Supply Chains, September 13, 2022.

[13] Project Team on the Evolution and Implementation of Ais, *The AI White Paper—Japan's National Strategy in the New Era of AI*, LDP Headquarters for the Promotion of Digital Society, April 2023.

Amazon Web Services, and Oracle, are joining a consortium focused on municipal AI governance, alongside the cities of Osaka, Tsukuba, Nagoya, and Yokosuka. Formed under the Japan-based Institute of Administrative Management, the group aims to develop guidelines for the use of AI in government services by April 2025. The initiative will be led by Yukihiko Okada from the University of Tsukuba's Institute of Systems and Information Engineering, while Japan's Ministry of Economy, Trade and Industry will participate as an observer, as will the Ministry of Internal Affairs and Communications, the Cabinet Office, and the Digital Agency.[14]

In February 2024, the Japan AI Safety Institute was established within the Information-technology Promotion Agency to focus on research related to safety evaluations, the development of standards, and the implementation of safety assessment methods. Additionally, the Institute is currently engaging with similar organizations in other countries, including AI Safety Institutes in the United States and the United Kingdom, to enhance international cooperation.[15] Since its foundation, the Japan Institute has released a «Guide to Evaluation Perspectives on AI Safety»[16] and a «Guide to Red Teaming Methodology on AI Safety.»[17]

In August 2024, following efforts in Europe and the United States to strengthen AI regulations, the Japanese government started discussions on potential laws and regulations for generative AI developers. Prime Minister Fumio Kishida has stated that the government aims to create a «flexible framework» that can keep pace with rapid advancements in technology and business and aligns with international guidelines, particularly the Hiroshima AI Process.[18] Last, to keep fostering and leading international cooperation, Japan's Prime Minister Fumio Kishida introduced at the OECD 2024 Ministerial Council Meeting in Paris the Hiroshima AI Process Friends Group, which includes participation from 49 countries and regions. The Group focuses on implementing the International Guiding Principles and the Code of Conduct to address the risks associated with generative AI,

---

[14] Akira Oikawa, Microsoft, Google partner with Japanese cities on AI rules, Asia Nikkei, April 30, 2024.

[15] EU-Japan Centre for Industrial Cooperation, *Launch of AI Safety Institute*, EU-Japan Centre for Industrial Cooperation, consulted on June 10, 2025. See also *Launch of AI Safety Institute*, Joint Release with Cabinet Office and Information-technology Promotion Agency, February 14, 2024.

[16] Japan AI Safety Institute, *Guide to Evaluation Perspectives on AI Safety* (Version 1.01), September 25, 2024.

[17] Japan AI Safety Institute, *Guide to Red Teaming Methodology on AI Safety* (Version 1.00), September 25, 2024.

[18] Georgia Adamson, *What Potential New Regulations Mean for Japan's AI Strategy*, Japan up close, September 26, 2024.

while fostering cooperation to ensure that people worldwide can benefit from safe, secure, and trustworthy AI technologies.[19]

6.1.3 United Kingdom

Another nation that opted to postpone implementing AI-specific horizontal regulations is the United Kingdom, which preferred adapting existing laws or interpreting a set of principles within a principle-based framework with the declared purpose of fostering innovation and ensuring AI was used to its full potential to benefit UK economy and society.

In 2021, the UK introduced an ethics, transparency, and accountability framework for automated decision-making, consisting of seven points to aid government departments in the safe, sustainable, and ethical use of automated or algorithmic systems.[20] Subsequently, several initiatives were launched by the Government: an investment exceeding £2.5 billion in AI since 2014, £110 million allocated for an AI Tech Missions Fund, and an additional £900 million designated to establish a new AI Research Resource and develop an exascale supercomputer for large AI models. This is further supported by an £8 million AI Global Talent Network and £117 million of existing funds aimed at creating hundreds of new PhD positions for AI researchers.[21] Also, in 2023 the UK launched the world's first AI Safety Institute, tasked with monitoring the fast-moving landscape of AI development, evaluating the risks AI poses to national security and public welfare and advancing the field of systemic safety to improve societal resilience.[22]

In 2023, the UK Government released a pro-innovation whitepaper on AI regulation,[23] explicitly detailing plans for implementing a pro-innovation approach to AI regulation and clarifying that they did not plan to enact new laws. In fact, the government was concerned that premature legisla-

---

[19] Speech by Prime Minister KISHIDA Fumio at the Generative AI Side Event, May 2, 2024.

[20] Department for Science, Innovation and Technology, Centre for Data Ethics and Innovation, Cabinet Office and Office for Artificial Intelligence, *Guidance Ethics, Transparency and Accountability Framework for Automated Decision-Making—Guidance for public sector organizations on how to use automated or algorithmic decision-making systems in a safe, sustainable and ethical way*, 2021.

[21] Secretary of State for Science, Innovation and Technology by Command of His Majesty, *A pro-innovation approach to AI regulation, Policy Paper*, presented to Parliament on March 29, 2023.

[22] The AI Safety Institute is a directorate of the UK Department for Science, Innovation, and Technology.

[23] Secretary of State for Science, Innovation and Technology by Command of His Majesty, *A pro-innovation approach to AI regulation, Policy Paper*, presented to Parliament on March 29, 2023.

tion could impose unnecessary constraints on businesses. The framework was thus underpinned by five principles to guide and inform the responsible development and use of AI in all sectors of the economy: safety, security and robustness; appropriate transparency and explainability; fairness; accountability and governance; contestability and redress; and it is aimed at being flexible and quickly adapted to new technologies or learnings from experience. However, these principles were not codified into law to avoid stifling innovation with new strict and burdensome legislative requirements on businesses.

This approach was chosen for its capability of adapting swiftly and appropriately to future technological developments. In fact, the principles have been introduced on a non-statutory basis and executed by existing regulators, which possess domain-specific expertise that allows them to adapt the principles' implementation to the particular contexts in which AI is applied.

However, a similar strategy was already employed by regulators even prior to the release of the whitepaper: for instance, in 2022, the MHRA (Medicines and Healthcare Products Regulatory Agency) issued a roadmap providing guidance on the requirements for AI and software used in medical devices, while concurrently working with other regulators to maintain regulatory coherence for AI through a formal network dedicated to AI and digital regulations in the health sector.[24] In this framework, a few case studies were presented to provide further guidance and indications, such as the example of automated healthcare triage systems. Initially, its adaptivity was examined, which referred to its capability to predict patient conditions based on pathology, treatment, and risk factors by analyzing medical datasets, patient records, and real-time health data. Next, its autonomy was defined as its ability to generate information about the probable causes of a patient's symptoms and suggest potential interventions and treatments, either for a healthcare professional or directly for a patient. Finally, the AI-related regulatory implications were highlighted, such as the unclear liability if an AI triage system gives incorrect medical advice, leading to adverse health outcomes for a patient and affecting their ability to seek redress.

Furthermore, in October 2023, the MHRA announced its new regulatory sandbox, the AI-Airlock, which was launched in May 2024 to assist in the development and deployment of software and AI medical devices, safely providing patients with earlier access to cutting edge innovations that

---

[24] AI and Digital Regulations Service, Care Quality Commission, Health Research Authority, Medicines and Healthcare Products Regulatory Agency, National Institute for Health and Care Excellence, 2023.

improve care.[25] Later, in 2024, the MHRA announced also its strategic approach to AI, welcoming the Government's pro-innovation white paper, and aiming to incorporate AI and technology into the healthcare system for faster, simpler, and fairer healthcare, while ensuring regulatory decisions are evidence-based. Last, the MHRA announced that they had been working on implementing a regulatory reform program for AI-driven medical devices, including risk proportionate regulation of AI as a medical device.[26]

Although the method chosen by UK offers significant flexibility, it also raises concerns about potential inconsistencies and divergences among different regulatory pathways; thus, strong central oversight and coordination are necessary to mitigate this risk. However, Sir Keif Starmer, Prime Minister of the United Kingdom since July 2024, has indicated that his administration will move UK toward a more regulation-friendly approach. Moreover, the UK technology secretary Peter Kyle, endorsed scientific and technological cooperation with the EU, in a change of tone from the previous government which highlighted the benefits of regulatory divergence from Brussels to capitalize on Brexit, indicating that the UK will bring in legislation to safeguard against the risks of AI in 2025. According to Kyle, the proposed AI bill would focus on making the UK's voluntary agreements on AI testing legally binding on the leading developers; focus exclusively on «frontier» models; and turn the UK's AI Safety Institute into an arms-length government body, giving it «the independence to act fully in the interests of British citizens,» thus separating it from the directorate of the Department for Science, Innovation and Technology.[27] Last, Kyle also launched a Regulatory Innovation Office to support healthcare AI innovations and reduce the burden of red tape and speed up access to new technologies that improve daily lives, from AI in healthcare to emergency delivery drones.[28]

In January 2025, UK government introduced the AI Opportunities Action Plan, detailing 50 measures to accelerate AI development and adoption.

---

[25] Medicines and Healthcare products Regulatory Agency, *MHRA to launch the AI-Airlock, a new regulatory sandbox for AI developers,* UK Government, published on 30th October 2023; Medicines and Healthcare products Regulatory Agency, *MHRA launches AI Airlock to address challenges for regulating medical devices that use Artificial Intelligence,* UK Government, published on May 9, 2024.

[26] Medicines and Healthcare products Regulatory Agency, *MHRA's AI regulatory strategy ensures patient safety and industry innovation into 2030,* UK Government, published on April 30, 2024.

[27] *UK will legislate against AI risks in next year, pledges Kyle,* Financial Times, November 7, 2024.

[28] Department for Science, Innovation and Technology and The Rt Hon Peter Kyle MP, *Game-changing tech to reach the public faster as dedicated new unit launched to curb red tape,* UK Government, published on October 8, 2024.

Key initiatives include: (1) forging new AI Growth Zones to speed up planning proposals and build more AI infrastructure; (2) increasing the public
compute capacity by twentyfold and developing a new supercomputer; (3)
creating a new National Data Library to safely and securely unlock the value
of public data and support AI development; and (4) establishing a dedicated
AI Energy Council chaired by the Science and Energy Secretaries, to work
with energy companies to understand the energy demands and challenges
which will fuel the technology's development. Also, the Action Plan promotes multiple pro-innovation regulatory initiatives.

6.2 Other pioneer countries leading in AI regulation and international policy shaping

Not all nations at the forefront of developing new AI governance are part
of the G7, even if the European risk-based approach strongly influenced
some of them, such as South Korea, Brazil, and Thailand, which opted to
regulate AI horizontally with specific measures for high-risk systems. In
contrast, countries like China and India are introducing numerous regulations targeting specific AI applications, with the clear purpose of fostering
innovation and avoid potential inhibitions coming from the burden of heavy
regulations.

6.2.1 South Korea

Similar to the European Union, South Korea chose to regulate AI horizontally. The «Basic Act on the Development of Artificial Intelligence and the
Establishment of Trust,» also known as the «AI Basic Act,» was passed into
law by the National Assembly in December 2024 and is expected to take
effect starting in January 2026. This law was strongly influenced by the EU
AI Act, as it adopts a risk-based framework emphasizing human oversight
and transparency. However, in contrast with the AI Act, the AI Basic Act
is less prescriptive and allows for more flexibility. High-impact AI applications in crucial domains like healthcare and employment, for instance, are
bound by more stringent requirements, such as providing user notifications,
incorporating human oversight, and implementing risk management plans.
The AI Basic Act does not, however, forbid any particular kind of AI system, in contrast to the EU AI Act.

South Korea, however, had already been working on national plans for AI
safe and effective implementation for a long time. In 2019, for instance, the

Ministry of Science and ICT issued the National Strategy on AI,[29] stating the aspiration for the country to rank third in global digital competitiveness by 2030, to generate 455 trillion Korean won in economic surplus through AI, and to join the top ten countries in terms of quality of life. Overall, South Korea's digital strategies align closely with the OECD AI Principles and their five policy recommendations.

As part of the Korean Digital New Deal, South Korea is also enhancing its digital ecosystem through data and products, including building a data dam by obtaining high-quality data for AI training and making public data available.[30] Additionally, South Korea is encouraging AI implementation by utilizing the extensive data in the public sector to aid the creation of AI services. For instance, the South Korean government has been offering companies experimental labs equipped with data, computing power, and memory to assist with medical imagery analysis, aiming to improve medical services in the military or support Covid-19 epidemiological research.[31] Also, South Korea went even further, recognizing the importance of AI education and literacy and offering foundational education programs to all citizens, including operating software schools and AI-centric institutions, along with providing online educational materials.[32]

Being aware of the critical relationship between privacy laws and AI regulation, the country amended its three key privacy laws in 2020 to encourage data use and align with the GDPR,[33] subsequently announcing new amendments around automated decision-making systems (including those powered by AI) to the Enforcement Decree of Korea's Personal Information Protection Act in 2024.[34] Also, the country showed a strong interest in the use of synthetic data, to the point that May 2024 the Personal Information Protection Committee of the Republic of Korea (PIPC) announced the establishment of five types of synthetic data generation reference models[35] aimed at aiding private researchers and companies in the generation and

---

[29] Government of the Republic of Korea, National Strategy for Artificial Intelligence, 2019.

[30] Ministry of Science and ICT, *Press Release: The Digital New Deal Is to Lead Digital Transition in the World after Covid-19*, July 15, 2020.

[31] Kyunghee Song, *Korea is leading an exemplary AI transition. Here's how*, OECD. AI Policy Observatory, March 10, 2022.

[32] *Ibid.*

[33] DLG law, *Data Protection Law in South Korea*, DGL website, 2025.

[34] Personal Information Protection Commission, *Press Release: In the era of artificial intelligence (AI), personal information safety devices are implemented*, March 6, 2024.

[35] Personal Information Protection Commission, Synthetic data were defined as newly created virtual data with statistical characteristics similar to actual data, so that similar results can be obtained from actual data analysis.

utilization of synthetic data for machine learning and artificial intelligence development.[36] This initiative was designed to facilitate the safe and proper creation of synthetic data, which are promoted as privacy-enhancing technologies because of their advantage of being able to be used without legal restrictions on personal information. These five types of synthetic data modeled also encompass a variety of data types, including oral images and blood sugar measurement information. The process of creating these models involved a comprehensive four-step methodology, which includes preparation, generation, verification of usefulness and safety, and utilization. All these steps ensure that the synthetic data maintains the statistical characteristics of actual data while safeguarding personal information.

Last, in April 2024, the Office of the President of South Korea released a statement[37] announcing ambitious plans for South Korea to become a global leader by 2030 in three key technological areas, through: the AI-Semiconductor Initiative, which will require intensive investment in AI models, chips, and hardware/software ecosystems to develop the next-generation AI technologies, low-power AI processors, and advanced packaging technologies; the Advanced Biotechnology Initiative, which includes nurturing digital biotechnology, promoting bio-manufacturing innovation, and developing advanced medical technologies to improve public health and tackle global challenges; and the Quantum Technology Initiative, focused on securing core quantum technologies, developing quantum engineering capabilities, and pioneering new quantum applications and AI.

6.2.2 Brazil

Brazil is another example of a country opting to regulate AI with a comprehensive law, instead of issuing multiple specific regulations. In May 2023, the Senate introduced Bill No. 2338 of 2023,[38] which governs the use of Artificial Intelligence, including consumer protection regulation, aiming to replace several prior bills that had already established principles for AI regulation. Heavily influenced by the European Union's AI Act, the bill takes a risk-based approach and sets out specific requirements for high-risk sys-

---

[36] Personal Information Protection Commission, *Press Release Personal Information Protection Commission Releases «Synthetic Data Creation Reference Model» for Researchers and Companies to Reference*, March 30, 2024.

[37] Office of the President Republic of Korea, Leap to Global Top 3 in AI—Semiconductor, Advanced Biotechnology & Quantum Technology, April 26, 2024.

[38] Bill No. 2338 of 2023 regulating the use of Artificial Intelligence, including algorithm design and technical standards, introduced in the Senate of Brazil on May 3, 2023 and approved on December 10, 2024.

tems, such as mandatory algorithmic impact assessments, usage restrictions, and penalties for violations. Also, the bill aligns with Brazil's General Data Protection Law (LGPD) to safeguard privacy and includes provisions for establishing a new authority to oversee and enforce AI regulations.

Moreover, in 2024, the Brazil Ministry of Science, Technology, and Innovations published the Artificial Intelligence (AI) Plan (PBIA) 2024-2028,[39] which calls for a planned investment of R$23 billion in AI over four years and for the creation of the National Center for Algorithmic Transparency and Trustworthy AI to ensure the transparency, security, and trustworthiness of AI systems in Brazil, including publication of guides on AI.

Brazil has also been taking part in several international efforts to foster cooperation on AI governance, especially in the LATAM region, which has been particularly active in issuing guidelines and recommendations to align with international AI principles and representing developing countries in the international forum. For example, in October 2023, the Santiago Declaration to Promote Ethical Artificial Intelligence in Latin America and the Caribbean[40] was issued in Chile, inspired by UNESCO's Recommendation on the Ethics of Artificial Intelligence. This declaration acknowledges that the possibilities and risks associated with AI require governments to anticipate and guide the creation and implementation of policies, plans, and strategies at national, regional, and international levels for the safe, ethical, and responsible design, development, and use of AI technology. It also calls for the establishment of effective national institutional frameworks, with a human rights focus, for proper AI management. The agreement brought together ministers and those in charge of digital and AI policies from Argentina, Brazil, Chile, Colombia, Costa Rica, Cuba, Dominican Republic, Ecuador, El Salvador, Guatemala, Honduras, Jamaica, Mexico, Peru, Paraguay, St. Lucia, St. Vincent and the Grenadines, Suriname, Uruguay, and Venezuela.

Another initiative that saw the involvement of Brazil was the 2024 Cartagena Declaration on AI governance,[41] signed in Colombia by representatives of 17 Latin American countries (Argentina, Brazil, Chile, Colombia, Costa Rica, Cuba, Curaçao, Ecuador, Guatemala, Guyana, Honduras, Panama, Paraguay, Peru, Dominican Republic, Suriname, and Uruguay), which ad-

---

[39] *Plano Brasileiro de Inteligência Artificial (PBIA) 2024-2028*, August 7, 2024.

[40] Cumbre Ministerial y de Altas Autoridades de América Latina y el Caribe, *Declaración De Santiago—Para promover una inteligencia artificial ética en América Latina y el Caribe*, Santiago de Chile, October 23–24, 2023.

[41] *Cartagena de Indias Declaration for Governance, the construction of AI ecosystems and the promotion of AI education in an Ethical and Responsible manner in Latin America and the Caribbean*, signed in Colombia on August 9, 2024.

heres to UNESCO's ethics recommendations from the AI for Good Global Summit and aligns to the OECD's AI recommendations. Furthermore, in Montevideo, Uruguay, the Ministers and High-Level Authorities representing the countries gathered at the Second Ministerial and High-Level Authorities Summit on the Ethics of Artificial Intelligence in Latin America and the Caribbean adopted the 2024 Montevideo Declaration,[42] reaffirming the region's commitment to development and deployment of AI based on the promotion, respect for and protection of human rights, fundamental freedoms, rule of law and democracy, focused on the human being, well-being and dignity of people, fostering innovation and ensuring inclusive and sustainable technological progress. Among other things, the declaration calls for a regional roadmap for AI with 5 priority lines of action to be implemented in the next 12 months on governance and regulation, talent and future of work, protection of vulnerable groups, environment, sustainability, and climate change and infrastructures.

Last, at the conclusion of the 19th G20 Summit in Rio de Janeiro, Brazil, leaders of the G20 countries adopted the 2024 Rio de Janeiro Declaration,[43] which, among other things, calls for strengthening of cooperation on the regulation of AI, taking into account the views of developing states, with the aim «to unlock the full potential of AI, equitably share its benefits, and mitigate risks» and the purpose of «work together to promote international cooperation and further discussions on international governance for AI, recognizing the need to incorporate the voices of developed and developing countries.»

6.2.3 Thailand

Thailand released its AI Ethics guidelines in 2021,[44] which outline ethical principles such as competitiveness and sustainable development, adherence to legal ethics and international standards, transparency and accountability, security and privacy, fairness, and reliability. Subsequently, in 2022, the Office of the National Digital Economy and Society Commission published

---

[42] *Declaration Of Montevideo—For the construction of a regional approach on the governance of Artificial Intelligence and its impacts on our society,* signed in Montevideo, Oriental Republic of Uruguay, by the Ministers and High-Level Authorities representing the countries gathered at the Second Ministerial and High-Level Authorities Summit on the Ethics of Artificial Intelligence in Latin America and the Caribbean, on October 3-4, 2024.

[43] *G20 Rio de Janeiro Leaders' Declaration,* November 18-19, 2024.

[44] Open Development Thailand, *Digital Thailand—AI Ethics Guideline,* 2021.

the draft Royal Decree on Artificial Intelligence System Service Business.[45] This framework was heavily inspired by the European Artificial Intelligence Act and adopted a risk-based approach by applying levels of regulatory scrutiny to AI system proportionally to the level of risk presented by the AI itself. For example, AI systems that pose unacceptable risks to public health, safety, or freedoms are generally prohibited, AI systems considered to be high-risk are subject to a conformity assessment, and AI systems considered to be limited-risk are subject to transparency requirements. Also, the draft decree explicitly bans AI systems that use subliminal methods to subtly influence human behavior, utilize social scoring, access sensitive personal information such as age or disabilities, or employ real-time remote biometric identification in public spaces.

In 2024, instead, the Ministry of Digital Economy and Society (DE) and the Electronic Transactions Development Agency (ETDA), through the AI Governance Center (AIGC), jointly issued the generative AI governance guidelines for organizations to establish a governance framework for the application of Generative AI at the enterprise level and to cover benefits and risks of generative AI.[46]

Last, in 2024, the Association of Southeast Asian Nations (ASEAN), which includes Brunei, Cambodia, Indonesia, Laos, Malaysia, Myanmar, Philippines, Singapore, Thailand, and Vietnam published the ASEAN Guide on AI Governance and Ethics,[47] offering practical advice for organizations in Southeast Asia aiming to design, develop, and deploy AI technologies for commercial and non-military or dual-use purposes. This guide adopts a risk-based approach, suggesting principles for AI use, promoting alignment within ASEAN, ensuring the interoperability of AI frameworks across different jurisdictions, and providing recommendations for national and regional initiatives to responsibly design, develop, and deploy AI systems.

6.2.4 China

During the Covid-19 pandemic, China demonstrated its capability to use AI for healthcare purposes. As the initial epicenter of the outbreak, the country

---

[45] *Key Concerns and Provisions in Thailand's Draft AI Regulation*, Tilleke and Gibbins, March 24, 2023.

[46] Electronic Transactions Development Agency, *Press Release DE and ETDA announce new Guideline! «Guidelines for applying Generative AI with good governance for organizations»*, October 30, 2024.

[47] ASEAN Secretariat, *ASEAN Guide on AI Governance and Ethics*, 2024.

leveraged AI to assist in restricting population movements, forecasting disease spread, and accelerating research for vaccine or treatment development. They were also successful in speeding up genome sequencing, enhancing diagnostic efficiency, conducting scanner analyses, managing maintenance, and deploying delivery robots.[48]

In China, the regulation of AI is primarily guided by the «Next Generation Artificial Intelligence Development Plan»[49] issued by the State Council of the People's Republic of China on July 8, 2017. This document, endorsed by both the Central Committee of the Chinese Communist Party and the State Council, urged Chinese governing bodies to foster AI development until 2030. Consequently, China opted for a distinct strategy compared to the European Union's one, and decided to focus on issuing specific laws for different AI applications instead of creating a horizontal framework. This method was probably also the first useful step toward establishing regulatory capacity before formulating comprehensive national AI law. According to the legislative plan released by the State Council in June 2023, such a law may be forthcoming.

By the end of 2024, China implemented rules on AI algorithms through three significant pieces of legislation.

The first regulation is the «Internet Information Service Algorithmic Recommendation Management Provisions,» which came into force in March 2022. It aims to address public dissatisfaction with e-commerce platforms' price discrimination and the algorithmic exploitation of delivery workers, among other problematic practices. In January 2023, China introduced the second regulation, known as «Internet Information Service Deep Synthesis Management Provisions,» which represents its initial attempt to regulate Generative AI to fight misinformation, especially following the surge of deepfake images. The third and most significant rule for general-purpose AI, the «Interim Measures for the Management of Generative Artificial Intelligence Services,» came into effect on August 15, 2023. These Interim GAI measures established several general requirements for the provision and use of generative AI services, mandating that providers and users adhere to core socialist values and avoid endangering national security and interests. AI-generated content must be labeled, effective anti-discrimination measures must be implemented, and personal data must be obtained with consent. Notably, these regulations focus more on preserving national security and social public interest than protecting individual rights, unlike many of

---

[48] Andy Hon Wai Chun, *In time of Coronavirus, China's investment in AI is paying off in a big way*, South China Morning Post, 2020.
[49] State Council Document No. 35.

the regulations issued by the other countries that were previously analyzed regulations.

In May 2024, China's Office of Central Cyberspace Affair Commission, the State Administration for Market Regulation, and the Ministry of Industry and Information Technology launched an ambitious three-year plan to establish the country as a global leader in AI and computing standards. The initiative, called the «Action Plan for Information Standard Construction (2024-2027),» focuses on strengthening standards in various cutting-edge technologies, such as artificial intelligence chips, generative AI, quantum information, and brain–computer interfaces.[50] Furthermore the Cyberspace Administration of China (CAC) released a draft regulation that aims to standardize the labeling of AI-generated synthetic content to protect national security and public interests.[51]

In December 2024, China's Ministry of Industry and Information Technology (MIIT) announced the establishment of a committee to develop AI standards, including for large language models and risk assessment, gathering representatives from major tech companies (like Baidu, Alibaba, Tencent, and Huawei) and universities. The formation of the AI standard committee is part of the strategy to establish at least 50 sets of AI standards by 2026, as announced by MIIT, together with three other government agencies, announced in July 2024.

However, China's efforts in regulating AI will only increase. According to a report submitted to the Standing Committee of the National People's Congress (NPC) during the third session of the 14th NPC, Chinese legislators aim to deliberate on 34 bills in 2025, with a focus on «intensify[ing] research on legislation in emerging sectors such as artificial intelligence, the digital economy, and big data, initiate the review and overhaul of laws in specific areas, and provide guidance to joint efforts in enacting interregional legislation.»[52] For example, new regulations and standards to govern the identification of AI-generated synthetic content were introduced in March 2025 by the Cyberspace Administration of China (CAC), alongside multiple ministries, with the goal of establishing a national standard for synthetic

---

[50] *Action plan to improve technology standards*, China Daily, May 30, 2024.

[51] China proposes new regulation on labeling AI-generated content, China Daily, September 14, 2024.

[52] Zhao Leji, Chairman of the Standing Committee of the National People's Congress Report On The Work Of The Standing Committee Of The National People's Congress—Delivered at the Third Session of the 14th National People's Congress, March 8, 2025.

content identification and cybersecurity guidelines for coding practices related to AI-generated content.[53]

However, China has also played a critical role in the international space. For instance, in October 2023, President Xi Jinping launched the Global AI Governance Initiative for Belt and Road Initiative countries, China's $1 trillion global infrastructure program which calls for the establishment of a testing and assessment system based on AI risk levels and the establishment and improvement of relevant laws, regulations and rules, with the aim to ensure personal privacy and data security. During this launch, President Xi Jinping stated that they «oppose using AI technologies to manipulate public opinion, spread disinformation, interfere in other countries' internal affairs, social systems and social order, or jeopardize the sovereignty of other states,» making clear the priorities of his administration related to AI governance.[54] Shortly after, China proposed a resolution on enhancing international cooperation and capacity building for AI, which was unanimously adopted by the 78th session of the UN General Assembly.[55] Furthermore, at the opening ceremony of the World Conference on Artificial Intelligence (WAIC) 2024 in Shanghai, the «Shanghai Declaration on Global AI Governance» was released.[56] The document calls for joint efforts to promote AI governance, international cooperation and healthy growth of the AI industry. However, it is a short document that lacks detail beyond a high-level emphasis on maintaining AI safety and promoting inclusive global cooperation. Similar to the Shanghai Declaration, both the Global AI Governance Initiative and the UN resolution lack clear specifics. Anyway, China promoted a UN-centered framework for AI governance and presented itself as a leader for the Global South, highlighting the significance of technology transfer and inclusive governance.[57] An explanation for this approach was developed by some experts who looked at recent developments and concluded that all the previous international efforts were mainly driven by Western countries,

---

[53] Cyberspace Administration of China, Notice on Issuing the Measures for Identifying Synthetic Content Generated by Artificial Intelligence, National Information Office Communication No. [2025] 2, March 14, 2025.

[54] Wang Cong, and Yin Yeping, «China launches Global AI Governance Initiative, offering an open approach in contrast to US blockade», *Global Times*, October 18, 2023.

[55] UNGA adopts China-proposed resolution to enhance int'l cooperation on AI capacity-building, Ministry of Foreign Affairs The People's Republic of China, July 4, 2024.

[56] Full text: Shanghai Declaration on Global AI Governance, Ministry of Foreign Affairs The People's Republic of China, July 4, 2024.

[57] Huw Roberts, *China's ambitions for global AI governance*, East Asia Forum, September 10, 2024.

sometimes to the exclusion of China.[58] Also, the release of ChatGPT and other large language models in late 2022 and early 2023 caught many, including China (which ended up banning ChatGPT), off guard, raising concerns about their potential impact on political stability. Furthermore, the release of these technologies shifted global AI governance discussions into the mainstream and probably increased China's concerns about the risks of being sidelined in global AI governance discussions. As a result, China's recent efforts to shape global AI governance may be interpreted as a reaction to these developments.[59] Moreover, in late 2023, Deepseek, a Chinese artificial intelligence company that develops large language models, was founded. In January 2025, Deepseek was already able to launch its own chatbot, a direct competitor of the US giant ChatGPT, marking a transformative moment in the AI race toward innovation, showing the US that, even if they had imposed trade restrictions on advanced Nvidia AI chips, China was still able to create cutting-edge innovation, thanks to smarter software and hardware optimization. Given that Deepseek, in contrast with its main competitors, is an open-source model, anybody can refine and use it without authorization or license agreements, including commercial firms and individual researchers. On one side, this encourages innovation and global AI development. On the other, this introduces significant risks, as Deepseek's company cannot monitor and restrict harmful applications.[60]

Last, China has also been looking at African countries with interest. For instance, at the China-Africa Internet Development and Cooperation Forum which focused on the cooperation between the African Union and China on data protection regulation within the domain of AI development, China and the African Union issued a joint statement aimed at enhancing cooperation in AI research and development, exchanging talent, and building a secure network and data protection framework. The statement aims to support the «Global Data Security Initiative,»[61] deepen cooperation in network and data security, and ensure the privacy and data security of in-

---

[58] *Ibid.*

[59] *Ibid.*

[60] Arun Rai, *How Deepseek is changing the AI landscape*, Georgia State News Hub, February 4, 2025.

[61] The Global Data Security Initiative is part of three Major Initiatives announced between 2021 and 2023 by the Chinese government—the others being the Global Development Initiative (GDI) and the Global Civilization Initiative (GCI), with the aim to help centralized China's position as a major world power within its reform of global security mechanisms. See Erik Green, Meia Nouwens and Veerle Nouwens, *The Global Security Initiative: China's International Policing Activities*, International Institute for Strategic Studies, October 24, 2024.

dividuals in AI research and application, being part of a broader effort to strengthen China-Africa cooperation in AI.[62]

6.2.5 India

India is one of the world's fastest-growing economies and of the most robust AI marketplace. The market is expected to reach $3,935.5 million by 2028, with a compound annual growth rate (CAGR) of 33.28% from 2023 to 2028. In 2022, the market was valued at $680 million. Furthermore, by 2025, AI might boost the country's GDP by around $500 billion. Furthermore, India is a prominent hub for innovation and research.[63] Moreover, India stands out as a center for research and innovation, and, since 2010, it has become the fourth-largest producer of scholarly papers related to AI, while ranking eighth globally in terms of AI patents filed in 2020.[64]

To align with all these developments, the Digital Personal Data Protection Act was introduced in 2023,[65] thus increasing requirements for specific AI developers. Also, India has been discussing for years how to better regulate AI. At first, in April 2023, the Ministry of Electronics and IT stated that while India was already adopting essential policies and infrastructure improvements to foster a strong AI industry, there were no plans to legislate its regulation.[66] This decision was likely made to safeguard the Indian start-ups entering the market, considering the sector a «significant and strategic» area for the country, and, sharing the same concerns of the UK's government, afraid that a burdensome regulation would hinder innovation. However, the administration subsequently hinted that the Proposed Digital India Act may also oversee AI high-risk systems.[67] This decision was later explained at the Confederation of Indian Industry's (CII) Annual Business Summit 2024, during which S. Krishnan, Secretary of the Ministry of Electronics and Information Technology (MeitY), stated that the Indian government was likely to apply a similar approach to the one used for drafting the Digital Personal Data Protection Act when creating

---

[62] *African Union: Adopted China-African Union Statement for Increased Cooperation in Artificial Intelligence*, Digital Policy Alert, April 3, 2024.

[63] Rashi Maheshwari, Aashika Jain, *Top AI Statistics and Trends*, Forbes Advisor, February 6, 2024.

[64] Husanjot Chahal, Sara Abdulla, Jonathan Murdick, and Ilya Rahkovsky, *Mapping India's AI Potential*, CSET Data Brief- Georgetown, March 2021.

[65] «The Digital Personal Data Protection Act» (No. 22 of 2023), 2023, *Gazette of India*, August 11, 2023.

[66] Manish Singh, *India opts against AI regulation*, Techcrunch, April 5, 2023.

[67] Sanhita Chauriha, *Explained: The Digital India Act 2023*, Vidhi Legal Policy, August 8, 2023.

regulations for AI. Krishnan also noted that India's delay in regulating AI would actually constitute an advantage, as it would open to the possibility of learning from the experiences and mistakes of other countries.[68] All the following initiatives to regulate AI undertaken by the government after this statement suggest that the main priority of the administration still remains to foster innovation in the country while providing a legal and ethical framework to ensure a trustworthy use of AI without imposing an overly burdensome regulation. For this reason, many options have been explored to ensure this aim.[69] In January 2025, a Report on AI Governance Guidelines Development, drafted by the Ministry of Electronics and IT, was released for public consultation.[70]

In general, the healthcare sector in India is currently facing significant challenges, including low public spending (just 1% of GDP) and a high burden of out-of-pocket expenses (71%). Key health indicators are concerning, such as a high prevalence of anemia among young women (56%), elevated infant mortality rates (47 per 1,000 live births), and a high maternal mortality rate (212 per 100,000 live births). In comparison to neighboring countries like Bangladesh and Sri Lanka, India lags behind in terms of overall health outcomes. The situation is particularly complex for the low-income population, who struggle to afford the high costs of healthcare services provided by the private sector, which currently caters to 78% of outpatients and 60% of inpatients.[71] In this context, the Indian Government recognized the poten-

---

[68] Soumyarendra Barik, *India will regulate AI, but not at the cost of innovation: Govt official*, Indian Express, May 17, 2024.

[69] For example, the Ministry of Electronics and Information Technology (MeitY) worked on a voluntary code of conduct and ethics for companies to follow in relation to the training, deployment, commercial sale and rectification of misuse of LLMs and AI platforms, see Aashish Aryan, *MeitY readying voluntary ethics code for AI firms*, The Economic Times, November 18, 2024; also, Minister of State for Electronics and Information Technology, Jitin Prasada, informed the Parliament that the government constituted an Advisory Group for India specific Regulatory AI framework with stakeholders including government officials, industry leaders, and academicians. Its mandate is to promote innovation and ensure adequate guardrails to protect common citizens against the possible misuse and user harms. Specifically, the terms of reference of the Advisory Group includes creating contextualized ethical guidelines which are adaptable in India and promote development of trustworthy, fair, and inclusive AI. The Government also organized the Global IndiaAI Summit and GPAI Summit in July 2024 and December 2023 where various stakeholders from government, industry, and academia engaged in discussions and deliberations for development of AI-based solutions in a safe and trusted manner, see Government Of India, Ministry Of Electronics And Information Technology, Lok Sabha, Unstarred Question No. 1484, To Be Answered.

[70] Ministry of Electronics and IT, Subcommittee on AI Governance and Guidelines Development, *Report on AI Governance Guidelines Development*, January 6, 2025.

[71] Planning Commission of India, Health Division. Report of the Steering Com-

tial of AI in healthcare early on. In June 2018, NITI Aayog, the National Institution for Transforming India and the primary public policy think tank for the Government of India, released the National Strategy for Artificial Intelligence #AIForAll. This discussion paper clearly outlined the government's role in fostering a research ecosystem, encouraging AI adoption, and addressing AI-related skill gaps among the population. The document also identified five key sectors likely to benefit significantly from AI: agriculture; education; smart cities and infrastructure; smart mobility and transportation; and healthcare, particularly regarding access to cure and affordability of quality medical services. The paper noted also that integrating AI into healthcare may mitigate the high barriers to accessing health facilities, especially in rural regions plagued by poor connectivity and a shortage of healthcare professionals. In this sense, potential applications of AI may include AI-driven diagnostics, personalized treatments, early detection of pandemics, and improved imaging diagnostics. However, the paper identified key obstacles to achieving these objectives, including the limited widespread expertise in AI research and application; the lack of robust data ecosystems; high costs and low awareness for AI adoption; concerns about privacy and security, including no formal regulations on data anonymization; and the absence of a collaborative approach to AI deployment and utilization.

Among various initiatives to reform the healthcare system, the Indian government seeks to use technology to enhance healthcare services also through the National eHealth Authority (NeHA), which plans for eHealth adoption, sets standards and frameworks for the health sector, and implements electronic health exchanges for interoperability.[72] The Integrated Health Information Program (IHIP)[73] aims instead to provide Electronic Health Records (EHR) to all Indian citizens and ensure interoperability with existing EHR/EMRs, along with establishing the Electronic Health Record Standards for India.[74]

---

mittee on Health for the 12th Five-Year Plan, New Delhi: Health Division, Planning Commission of India, 2012. See also Sunil Kumar Srivastava, «Adoption of electronic health records: a roadmap for India», *Healthcare Informatics Research*, 22(4), 2016, pp. 261-269.

[72] Manisha Wadhwa, National eHealth Authority (NeHA), «CSD Working Paper Series: Towards a New Indian Model of Information and Communications, Technology-Led Growth and Development», *ICT India Working Paper #29*, April 2020.

[73] The Integrated Health Information Platform by the Ministry of Health and Family Welfare provides a list of national health programs at https://ihip.mohfw.gov.in/#!/.

[74] «Digitalisation of healthcare data through electronic health records will be the next startup boom», *Economic Times*, February 23, 2024.

Additionally, in recent years, both public and private sectors in India have developed AI-powered tools that enhance the efficiency of health and security services. These innovations include the «MyGov Corona Helpdesk,» a chatbot designed to increase Covid-19 awareness and ready India for its challenges, and «e-Paarvai,» an advanced AI-powered mobile application created to detect cataracts and address the shortage of ophthalmologists.[75]

6.3 Unleashing the potential of AI in African countries

African nations are aiming to harness AI technologies to address numerous social and economic challenges, as well as to support the achievement of national development goals while enhancing healthcare access in rural areas.[76] For example, globally, Africa has the highest burden of infectious diseases: Furthermore, of the estimated 10 million deaths per year resulting from infectious diseases, the majority occur in Africa. Infectious diseases exert adverse clinical and economic impacts on the continent and annually account for over 227 million years of health life lost and produce an annual productivity loss of over $800 billion.[77] However, the success achieved by AI platforms in the Covid-19 pandemic (e.g., rapid collection and real-time dissemination of data and vaccine development) can be adapted to combat infectious diseases on the continent.[78]

Moreover, AI applications are making significant strides on the continent, particularly in healthcare. For example, CareAi (a blockchain-anchored AI system funded by the European Commission) is employed in various African countries to rapidly diagnose infectious diseases such as malaria, typhoid fever, and tuberculosis within seconds.[79] In a different context, Rwanda began utilizing drones in 2016 to deliver critical blood supplies to remote rural areas, and the country is now recognized for its role in advancing global drone legislation.[80] Additionally, Mauritius was the pioneer among African nations

[75] Jibu Elias, *AI for All: How India is carving its own path in the global AI race*, OECD.AI Policy Observatory, January 30, 2023.

[76] Jonathan Guo and Bin Li, «The application of medical artificial intelligence technology in rural areas of developing countries», *Health Equity*, 2018.

[77] Idemudia Otaigbe, «Scaling up artificial intelligence to curb infectious diseases in Africa», *Frontiers in Digital Health*, 4, 2022.

[78] Lian Wang *et al.*, «Artificial intelligence for COVID-19: a systematic review», *Frontiers in Medicine*, 8, 1-15, 2021.

[79] Ahmad Abu el-Hamd, «Care AI a European Commission new technology that can transform Africa's Health Care System», *The Middle East Observer*, 2018.

[80] Noel Stierlin, Martin Risch, and Lorenz Risch, «Current Advancements in Drone Technology for Medical Sample Transportation», *Logistics*, 8(4), 2024, p. 104.

to release a developmental roadmap on AI. This plan also expressed their intention to incorporate artificial intelligence into healthcare services, aiming to create patient databases (with consent) for testing AI applications such as deep genomic analysis, DNA sequencing, and personalized medicine.[81]

Surely, AI holds immense promise for transforming healthcare in Africa. However, while the continent plays a key role in the global AI supply chain, particularly in the early production stages, many AI solutions are developed abroad and can introduce substantial risks when implemented in African countries. For instance, the first challenge pertains to the lack of large clinical datasets for training AI models: with the low level of digitization and electronic medical record use across Africa, there is a paucity of locally generated useful data that are important for building AI systems. Also, algorithmic bias has a more pronounced effect when AI applications are introduced to the African setting, given that most AI applications are being developed outside Africa and most datasets available are from people who differ from Africans physiologically.[82] For instance, facial recognition systems have been shown to misidentify African individuals, leading to serious human rights violations.[83] Other issues affecting AI use in Africa relate to the costs of implementing and maintaining AI systems and the cost or lack of infrastructure.[84]

Overall, the policy landscape for AI in Africa is still evolving. Countries like Mauritius, Egypt, and Rwanda have already published national AI strategies, yet numerous challenges remain. Data governance is one such hurdle, with the African Union Commission pushing for progress through the Continental Data Policy Framework.[85] Additionally, several nations have recently enacted privacy laws, particularly following the implementation of the GDPR. However, these laws do not yet fully address all the complexities associated with AI usage and fall short of the GDPR's comprehensive protections. Consequently, European personal data processed in Africa often enjoys stronger safeguards against misuse than the personal data of Africans.[86] Other policy measures and challenges that African

---

[81] *Mauritius Artificial Intelligence Strategy*, A report from the working group on Artificial Intelligence, 2018.

[82] Ayomide Owoyemi *et al.*, «Artificial intelligence for healthcare in Africa», *Frontiers in Digital Health*, 2(6), 2020.

[83] Rachel Adams, «AI in Africa: key concerns and policy considerations for the future of the continent», *Policy Brief*, No. 8. Berlin: APRI, 2022.

[84] Eric David Halsey, «What does AI actually cost?», *Medium*, 2017.

[85] AU Data Policy Framework, African Union, 2022.

[86] Benedikt Erforth, *Data Extraction, Data Governance and Africa–Europe Cooperation: A Research Agenda*, Working Paper 14, Megatrends Afrika, July 2024.

governments are prioritizing include developing infrastructure, enhancing workforce skills, and fostering regional cooperation to prepare the continent for effective AI deployment.

For example, Smart Africa represents a strong commitment by African leaders to boost sustainable socioeconomic development through affordable Broadband and ICT.[87] Initiated at the Transform Africa Summit in 2013, the Smart Africa Manifesto was initially embraced by seven nations. By January 2014, it gained endorsement from all African Union Heads of State, extending its reach to all 53 African countries. Now, the Smart Africa Alliance includes 40 countries, representing over one billion people. In 2021, SmartAfrica unveiled the «Blueprint for AI in Africa,» outlining strategies for AI governance and local development on the continent. This document highlighted various applications of AI in healthcare, including advantages such as empowering medical professionals; enhancing limited personnel and facilities to focus more on clinical outcomes by automating initial processing, triage, diagnoses, and post-care follow-ups; improving diagnostics through computer vision in medical imaging; and expanding healthcare access using chatbots in local languages to reach millions of remote individuals via smartphone cameras for pre-screening or diagnoses. Notably, Babyl, a digital healthcare provider in Rwanda, was cited as a successful case, as it collaborated with the national healthcare system to leverage AI and machine learning to offer medical advice and schedule appointments through mobile apps. Also, although not yet part of their services, the company aims to incorporate AI-based medical diagnosis in the future.[88]

6.3.1 South Africa

South Africa is facing social and economic challenges, alongside a significant disease burden and health equality issues.[89] Even though AI is currently not regulated, its potential to further the nation's development, including in the healthcare industry, has been acknowledged and investigated. For example, in 2016, the Academy of Science of South Africa initiated a consensus study on the ethical, legal, and social aspects of genetics and genomics in the country.[90] The goal was to produce a document to aid the national Depart-

---

[87] All information on the official website https://smartafrica.org/

[88] *Ibid.*, p. 51.

[89] Katusha de Villiers, «Bridging the health inequality gap: an examination of South Africa's social innovation in health landscape», *Infectious Diseases of Poverty*, 10(19), 2021.

[90] Michael S. Pepper *et al.*, «ASSAf consensus study on the ethical, legal and social implications of genetics and genomics in South Africa», *South African Journal*

ments of Health and Science and Technology in creating legislation, regulations, and guidelines regarding human genetics and the human genome.

In 2019, the South African government established the Centre for the Fourth Industrial Revolution in South Africa, aimed at promoting smart regulation for new technologies like AI and blockchain. This effort involved collaboration between government regulators and the technology industry in South Africa and also led to the formation of a Presidential Commission on the Fourth Industrial Revolution. Later in 2020, the Presidential Commission released a report stating that «the high-technology industries in which the country is significantly underperforming include artificial intelligence, blockchain, virtual/augmented reality simulation environments, automatic data-processing machines, electrical and electronic goods, biotechnologies, storage/transmission, advanced materials, advanced sensor platforms, as well as medicinal products and pharmaceuticals.»[91] In 2022, the Artificial Intelligence Institute of South Africa and several AI hubs were launched following the report's recommendations. Last, in 2024, also the South Africa National Artificial Intelligence Policy Framework[92] was launched, which focuses on key pillars that will shape South Africa's AI landscape.

Interestingly, in 2021, South Africa became the first country to grant a patent to an AI inventor.[93] In contrast, other countries as the United States, the United Kingdom, and the European Union rejected similar applications. However, the rationale behind South Africa's divergent decision might be found by examining key policy documents such as the «Intellectual Property Policy of the Republic of South Africa Phase I,»[94] the Department of Science and Technology's «White Paper on Science, Technology, and Innovation,»[95] and the report of the Presidential Commission on the 4th Industrial Revolution.[96] These documents collectively suggest that the South African government aims to boost innovation to address socio-economic challenges, including low levels of innovation, insufficient funding, and inadequate

---

of Science, 2018.

[91] Department of Communications and Digital Technologies Notice 591 Of 2020.

[92] Department Communication and Digital Technologies—Republic of South Africa, South Africa National Artificial Intelligence Policy Framework, 2024,

[93] Patent Application No. PCT/IB2019/057809, filed on September 17, 2019.

[94] Intellectual Property Policy of The Republic of South Africa Phase I, Department Trade and Industry—Republic of South Africa, 2018.

[95] Department Science and Technology—Republic of South Africa, *White Paper on Science, Technology and Innovation*, June 2019.

[96] Report of the Presidential Commission on the 4th Industrial Revolution, Government Gazette, October 23, 2020, No. 43834.

infrastructure. Thus, awarding patents to AI may form part of a broader strategy to transform the nation into a more innovative society.[97]

6.3.2 Nigeria

Nigeria is today a fast-growing technology start-up hub (having secured 25% of the $1.3 billion funding for African tech start-ups in 2021). Because of this, the country is highly engaged with AI applications.[98] Already in 2020, the Nigerian Communications Commission published the «Ethical and Societal Impact of AI,»[99] a research paper that examines the societal implications and ethical considerations related to the widespread deployment of AI, particularly highlighting its potential transformative effects on local healthcare. Additionally, in 2023, the country enhanced its data framework by enacting the Nigerian Data Protection Act, which sets forth principles and guidelines for the legal collection, processing, and storage of data.

Even if Nigeria has chosen to not yet regulate AI, it is currently developing a National AI strategy aimed at fostering ethical and inclusive AI innovation, with the purpose of enhancing welfare and expanding opportunities for its citizens.[100] The strategy outlines the creation of a «strong AI governance framework,» which includes a set of comprehensive National AI Principles that will define Nigeria's core values regarding AI development, deployment, and use; an AI Governance Regulatory Body responsible for providing clear guidance and enforcing ethical standards; a National AI Policy Framework that will establish definitions and guidelines for the responsible development, deployment, and use of AI; and a National AI Risk Management Framework that will set protocols for identifying, assessing, and mitigating potential safety and security risks related to AI systems.

Overall, Nigeria's strategy also employs a co-creation approach, which is designed to involve leading AI researchers of Nigerian heritage worldwide.[101]

---

[97] Meshandren Naidoo, «In a world first, South Africa grants patent to an artificial intelligence system», *Mail and Guardian*, August 11, 2021.

[98] Bosun Tijani, *Co-creating a National Artificial Intelligence Strategy for Nigeria*, 28th August 2023.

[99] *Ethical and Societal Impact of AI*, Nigerian Communications Commission, 2020.

[100] Federal Ministry of Communication, Innovation and Digital Economy, *Draft National Artificial Intelligence Strategy*, August 2024.

[101] Bosun Tijani, *op. cit.*

7   Conclusions

7.1 Final thoughts and recommendations

This book aims to study the applications of artificial intelligence in health-care and how law and policy intersect in this realm. We started by analyzing the definition of artificial intelligence, and the implications of reaching or not a consensus about what responsible use of AI means. Also, the unique and specific characteristics of AI use in the healthcare sector were represented in order to deepen the understanding of why this area requires distinct considerations, given its direct impact on human health, the importance of research of new treatments, and ethical and legal complexities linked with these features.

Our discussion moved later on to examine the ethical framework that usually guides AI development and deployment, which is based on the four principles of beneficence, non-maleficence, autonomy, and justice. We then deepen into the associated ethical issues of protecting autonomy, promoting human well-being, guaranteeing human safety and the public interest, ensuring transparency, explainability, and intelligibility, fostering responsibility and accountability, ensuring inclusiveness and equity, and promoting artificial intelligence that is responsive and sustainable.

Moving to the legal framework, we analyzed various legal concerns partly influenced by the ethical ones but differing from them primarily due to their enforceable nature and grouping them into safety and effectiveness, liability, data protection and privacy, cybersecurity, and intellectual property law. We discussed how, in some cases, AI enters areas already governed by regulations established before its emergence, while in others, specific provisions are needed to address the unique challenges and opportunities it presents.

Policymakers are facing unique challenges when it comes to AI regulation. On one side, local efforts are likely to compromise international ones and hinder the quest for reaching global harmonization in terms of definitions and approaches. Also, policymakers are challenged with the need to consider healthcare sector-specific needs when developing AI regulations, which often involves evaluating how to best apply the risk-based approach to each different AI use in very different settings, such as clinics or research ones. Other complexities are brought up in cases in which dynamic AI models and black boxes are used. In this context, ensuring accountability and liability even where there may be a significant loss of human control becomes extremely important, as well as ensuring transparency and explainability from AI systems providers and in the relationships between healthcare professionals and patients, while promoting AI literacy and education between healthcare professionals, patients, and institutions. Lastly, AI inventorship and AI environmental impact are issues that are receiving increasing attention from policymakers.

We later analyzed the regulatory frameworks in the European Union and the United States to highlight the distinct approaches to regulating AI in healthcare. We examined the European precautionary approach, which prioritizes safety and protection of fundamental rights through comprehensive frameworks like the General Data Protection Regulation (GDPR) and the AI Act. On the other hand, the United States adopted a market-driven approach, choosing to regulate AI with sector-specific laws to foster innovation and remove potential inhibitions to AI growth. In the final chapter of the book, we also looked at other approaches that could potentially be adopted in AI regulation, analyzing how other nations are developing AI laws to suit their own needs. We focused, each time, on frameworks that prioritize principles and effects on society, state control, promoting innovation, or public–private cooperation. In any case, we concluded that international cooperation remains critical to ensure harmonization of terminology and risk frameworks so that regulatory fragmentation can be reduced through global standards.

In the course of the book, we examined considerations regarding two fundamental needs: balancing the necessity of safeguarding individual privacy with the requirement of data sharing for research purposes and balancing the goal of fostering innovation with the need to ensure patient safety.

We have deeply analyzed the conflict between free data sharing for research purposes and privacy protection for personal data. On one side, we recognized that researchers looking for revolutionary medical insights need to be able to access and analyze large datasets, as they are the true enablers of innovation and will allow them to progress in scientific discovery. However, we cannot stress enough the importance of rights such as privacy and ensuring that individuals retain control over their health information.

Nowadays, developments in data analytics and new machine learning techniques raise significant concerns about patient identification and the potential exploitation of sensitive information, especially considering the increasing amount of digital health data collected and stored. Empowering research without compromising privacy rights can be achieved only through effective data governance frameworks, which should clearly define how data is collected, stored, accessed, and shared, covering anonymization and consent mechanisms, and emphasizing the importance of data security and ethical practices.

At the same time, emphasis should be put on the importance of promoting data-sharing policies, including incentives to motivate patients, institutions, and researchers to share data more openly. These policies should also encompass interoperability and standardization of data formats so that information is compatible across systems and countries. Lastly, public–private partnerships between public health agencies, research centers, and private companies should also be enabled as a tool to pool valuable resources and knowledge for collective progress.

However, another sensitive matter we examined is the promotion of innovation while guaranteeing patient safety, looking at the approaches adopted by the European Union and the United States, particularly California. In the previous chapters, we analyzed the particular concerns associated with the use of AI in healthcare and clinical practice, including algorithmic bias, data drift, and excessive dependence on automated systems. These issues, if left unaddressed, may have dangerous consequences, such as discrimination, adding barriers to healthcare access, and generating wrong outputs and recommendations. All these could endanger the rights of individuals. In practical terms, even while AI has enormous potential to transform healthcare, patient safety must not be sacrificed for these advantages.

As an example, in this evolving framework, AI is proving to be a tool that does not substitute humans, but enhances the work of professionals, while optimizing processes and redefining interactions between clinicians and patients. This evolution does not eliminate the need for clinical judgment, but, on the contrary, requires doctors to reinterpret their role by placing at the center the need to become critical mediators between algorithmic data and the specific context of the individual patient, taking into account during the care path all those aspects that complex predictive algorithms, such as social status, economic conditions, or the personality of the patient could otherwise overlook.

However, the issues of balancing the necessity of safeguarding individual privacy with the requirement of data sharing for research purposes and fostering innovation while ensuring patient safety can be addressed through careful policymaking and law drafting. Sometimes, we may think that law can be an obstacle to innovation and research, looking at it as a barrier more than as a tool. However, the law is, exactly like artificial intelligence, a human invention, and, as such, it can be redirected to address the most pressing needs or to boost human and technology capabilities. Law is not and should not be considered as an issue, but as the most powerful tool to ensure that progress is reached in an ethical and responsible way.

Looking ahead, striking the right balance between data sharing for research and privacy, or between innovation and safety, will be possible only through a multi-stakeholder approach that includes companies, patients, medical professionals, ethicists, legal professionals, and technologists in the policy-shaping and law-making processes. All actors in society must be equipped with the necessary AI literacy to successfully understand and interact with AI, but also to ensure that they are aware of the implications of regulating or not regulating AI and thus that they can effectively dialogue with policymakers and ensure that their voice is heard and their instances are taken into consideration in the law-making process. This type of stakeholder engagement is the essential premise needed in any strategy aimed at mitigating AI risks and ensuring that AI technologies are used ethically and safely. Only then will AI be optimized to its full potential to enhance healthcare outcomes while preserving individual and collective rights.